# BOAC
# AND THE GOLDEN AGE OF FLYING

## 1939 — 1974

# BURNT ASH PUBLISHING

First published in Great Britain in 2019
Reprinted in 2020

Burnt Ash Publishing
A Division of Burnt Ash Developments Limited
86-90 Paul Street
London
EC2A 4NE

Printed by Cambrian Printers, Aberystwyth, Wales
Typeset in Bookman Old Style

ISBN 978-1-9162161-0-5

A CIP Catalogue for this book is available from the British Library

## SUBSCRIBERS

We are indebted to our subscribers who generously supported this book.
Their names are shown in date order.

Keith Palmer
Alec Moore
David Nicholas
Paula Joyce
Christopher Le Breton
Bob Nipperess
Ron Beaumont
Louis Ferguson
Thomas Hilditch
Peter Adams
Robert Haddock
Nigel Maddock
Norman Giffin
Michael Couzens
Peter Brown
Richard Bartholomew
Jeremy Tweedie
Vikki Ball
Glenn Haley
David Hedges
Stefano Pagiola
Jacob Avon
Steve Carroll
Adrian Shaw
Keith Rennison
Geoffrey Weill
Jeremy Boyd
Rohn Dubler
Tonii Ford

Ian Cave
Andrew Gardner
Terry Ely
Paul Lonsdale
Jay Miranda
Martin Waters
Donald Johnston
Mark Saunders
Susan Katzban
Michael Spurling
Stewart Weller
Ray Howell
Marian Griffiths
Carol Jempson
Graeme Catnach
Richard Palmer
Nick Lloyd
Tony Allerton
Phil Hardyman
William Bird
Andrew Hawkins
Peter Hill
Jacob Drudge
Christopher Goodfellow
Mark Genet
Brian Soper
Paul Luscombe
Frederico Fiori
Robert Barnes

# CONTENTS

Far better travel by

B·O·A·C

# INTRODUCTION

*'The past is a foreign country: they do things differently there.'*

    – L P Hartley, *The Go Between*

One of my earliest memories is meeting my father at London Airport. In the days before the Oceanic Building opened, BOAC's Stratocruisers and Constellations taxied up to the northern perimeter and delivered their passengers onto a bare concrete apron in front of the control tower. If it was raining, uniformed ground staff held umbrellas above the passengers' heads and guided them to a bleak, prefabricated concrete building which served as the arrivals hall (not long before, it had been an army surplus tent).

Despite the sombreness of a half-built post-war airport, those scenes are burned into my memory and tinged with romance. In austere 1950s Britain the very idea of foreign travel, never mind being flown across the Atlantic Ocean, struck me as the height of sophistication and adventure. The passengers themselves, whose clothes were obviously expensive and possibly foreign, seemed to bring a bright shaft of colour into my dull English schoolboy's life.

The aircraft, then painted in an elegant white livery with thin blue horizontal stripes and the initials B.O.A.C. spelt out in enormous blue letters, shimmered in the heat haze of their exhausts. My father was a pilot, and when he emerged from the arrivals building in his blue and gold uniform with the breezy detachment that pilots always seemed to display when not actually flying, I thought that he was a god and that heaven probably looked something like this airport. Sixty years later, these memories flood back to me along with the sound of mighty piston engines and the smell of aviation fuel.

At a very early age the belief took root in my mind that BOAC, or to give it its full name, The British Overseas Airways Corporation, was something above the ordinary, and that idea has never entirely gone away. This book is a celebration of an iconic British airline and a brief period of history which is sometimes called the Golden Age of Flying.

CHAPTER ONE

## Flying's Golden Age

*'The journey not the arrival matters.'*
—Leonard Woolf, 1969

In 1963 Elizabeth Taylor and Richard Burton were the most famous couple in the world. Their love affair, which had become public during the filming of *Cleopatra*, caught society's imagination as no other romance would until Charles married Diana in 1981. Taylor and Burton would not finally be able to wed until 1964, as each needed time to divorce their spouses. In the interim, *Cleopatra*, which had taken four years to complete and endured massive cost overruns, would become a huge box office success. Shortly after it was released and at the very peak of their fame, the lovers would make one more film together before taking a two-year break.

*The VIPs,* written by Terrence Rattigan and directed by Anthony Asquith, cleverly mirrored the lives of its stars. Elizabeth Taylor

*Elizabeth Taylor and Richard Burton starred in the 1963 film, The VIPs*

played a famous model attempting to leave her husband for her lover, and Richard Burton played her millionaire husband. With an all-star cast and a huge budget, the producer and director needed a suitably glamorous location. In a different era, such a film might

In a dramatic moment of passion, Elizabeth Taylor has smashed a mirrored door as Richard Burton rushes to aid her.

Metro-Goldwyn-Mayer presents "THE V.I.P.s" in PANAVISION® and METROCOLOR

have been set on the Côte d'Azur or in a Venetian palazzo, but in the early 1960s the public's view of glamour was altogether different. *The VIPs* was set almost entirely in what is now the departure lounge of Terminal 3 at Heathrow Airport. Most of the filming was actually done at the MGM-British Studios in Borehamwood where a huge replica of Terminal 3 had been constructed. At the time it was the largest film set ever built in the UK.

Today, in an era of low-cost travel, strict airport security, and hid-

eously crowded terminals, such a location would rarely be used for anything but a crime drama or a gritty fly-on-the-wall documentary. But in a decade when barely one per cent of the world's population had ever flown, airports were thought to be

9

*The VIPs, written by Terrence Rattigan and directed by Anthony Asquith, cleverly mirrored the lives of its stars.*

the epitome of sophistication. This was the same decade that coined the term Jet Set to replace what had formerly been known as Café Society. In the West End of London, the play *Boeing Boeing*, a comedy based on the tangled love life of a bachelor who is simultaneously dating three stewardesses, filled theatres for seven years. At the beginning of the sixties, Federico Fellini's style-setting movie *La Dolce Vita* had featured long lingering shots of jets and their passengers coming and going from Rome's Ciampino airport.

Virtually all of the passengers depicted in *The VIPs* are wealthy and glamorous but the only other thing they have in common is that they are booked aboard the same flight to New York. The BOAC Boeing 707, surrounded by dense, swirling fog, remains stubbornly on the ground but still manages to look seductive. The 707 was designed to be sleek and sexy, in marked contrast to the lumpy, functional-looking planes we travel on today. It had neat, square windows, clean swept lines and its high frequency radio aerials had been cleverly crafted to extend like spears from the tail and wing. Its name even chimed with 007, the coded soubriquet of James Bond. Boeing's stylish design paid off; the 707 outsold its prosaic rival, the DC8, by a wide margin.

*The 707 was designed to be sleek and sexy, in marked contrast to the lumpy, functional-looking planes we travel on today.*

*The VIPs'* drama revolves around the problems, financial and emotional, that will be created in the lives of the passengers by their delayed departure. Throughout the film uniformed BOAC staff interact with the passengers, BOAC signs and logos are everywhere, and the camera regularly cuts to a misty shot of the aeroplane. The publicity value for BOAC must have been immense: the unmistakable message was that jets, especially *BOAC* jets, are the way that glamorous people travel.

*In Out of the Clouds, James Robertson Justice (right) played an irascible BOAC Captain.*

It was not the first time that BOAC had been featured in a movie. *Out of the Clouds,* is a 1955 British film directed by Basil Dearden and starring Anthony Steel, Robert Beatty and James Robertson Justice. An Ealing Studios production, it is a drama documentary that fairly accurately describes a day in the life of London Airport. The Ministry of Transport and Civil Aviation co-operated in the production of the film and BOAC starred as itself, with walk-on parts for British European Airways and Pan American. The film used one of Ealing Studios' largest ever sets to recreate the interior of the terminal building as well as the control tower. There are beautifully filmed flying scenes of Stratocruisers, Constellations, Viscounts and DC7s, as well as realistic cockpit and air traffic control scenes. A Stratocruiser returns to London with one engine shutdown and another overheating, which is entirely credible – the Wasp engine was notoriously unreliable and double-engine failures were not uncommon. James Robertson Justice plays an irascible captain, a role probably inspired by the real BOAC pilot, Captain O P Jones.

*Cone of Silence is a 1960 British film loosely based on the Comet ground stall accidents.*

*Cone of Silence* is a 1960 British drama film directed by Charles Frend starring Michael Craig, Peter Cushing, George Sanders, and Bernard Lee. It is loosely based on the Comet 1 ground stall incidents. The idea came from David Beaty's 1959 novel of the same name. Although the drama seems amateurish by today's standards, it sticks broadly to the facts. The fictional 'British Empire Airways' is plainly meant to be BOAC and the Comet-based 'Phoenix' jet airliner is represented by an Avro Ashton. The real aircraft was used when taxiing, but unfortunately an unconvincing model was used for the flying shots.

BOAC itself, throughout its relatively short history (1939–1974), had used its advertising and publicity to exploit the idea that air travel was for the elite. As a wholly state-owned organisation, they were obliged to do two things: the first was to 'fly the flag' for Britain across her far-flung but diminishing Empire, and the second was to act as a proving ground for the products of the British aircraft industry. Both these obligations weighed heavily on profits and the corporation often lost money as a result.

In the early 1960s BOAC had tried to cancel an order for the British-built Vickers VC10, preferring to operate an all-707 fleet. Their logic was simple; the 707 consumed less fuel per passenger and was therefore more profitable. Cancellation of the entire BOAC order would have spelled doom for the VC10, if not for the British Aircraft Corporation, which Vickers and several other aircraft manufacturers had recently been merged into. The airline came under considerable pressure from a government which considered the VC10 project to be vital to the national interest. In the end, a classic British compromise was reached: BOAC would reduce its order and operate a combined fleet of Boeing 707s and Vickers VC10s in roughly equal numbers.

*The rear-engined VC10 gave passengers a quieter and smoother ride (particularly in first class).*

*The VC10 became a design icon which has rarely been equalled.*

BOAC's advertising department, along with their advertising agents, Foote, Cone & Belding, went to work on selling the new aircraft to passengers. It was no small challenge: the 707, which the press had nicknamed, Queen of the Skies, had already gained considerable passenger appeal. BOAC, however, was helped by two assets: the first was that the rear-engined VC10 gave passengers a quieter and smoother ride (particularly in first class), and the second was that the new British plane was utterly beautiful. In contrast to the more angular 707, she had gorgeous curves and an especially graceful tail. In fact, there was almost no direction from which you could view her and not be impressed. Along with the Spitfire and Concorde, the VC10 became a design icon which has rarely been equalled.

A clever media campaign had worked. Passengers fell in love with the distinctive new British jet and insisted on flying in her, even if it meant delaying their departure. Before long, BOAC found that the iconic VC10 was actually making more money for them than the 707! In a glamorous age, a glamorous, beautiful plane had won out over basic economics.

# CHAPTER TWO

## In the Beginning

*'Small opportunities are often the beginning of great enterprises.'*
—Demosthenes (384 BC – 322 BC)

This book is more of a celebration than a history, but we cannot understand how BOAC became such an icon without first knowing something of its provenance.

The British were not early converts to flying and, as with so many other things, it would take a world war to change their minds. In October 1908, a group of senior army officers had gathered in a field in Farnborough to witness Britain's first official flight of a manned, heavier-than-air flying machine. It was designed, built and flown by Samuel Franklin Cody, a flamboyant American showman who had previously toured the music halls with a Wild West act. His unconventional background, eccentric dress and unreliable claims about his past may have been part of the reason that the generals were unimpressed – along with the fact that the first flight ended in a crash which severely damaged the machine! A year later the War Office decided to stop backing the development of heavier-than-air aircraft, convinced that they had little military value.

*Samuel Franklin Cody.*

Senior British officers remained sceptical about aircraft right up until the outbreak of war in 1914, preferring to put their faith in balloons for artillery spotting and observation. Opinions only changed once it became clear that barbed wire, trenches and modern weapons meant that the cavalry could no longer carry out their traditional role of reconnaissance. When that lesson had been fully taken on board, it caused a Damascene conversion and a scramble for the new-fangled aeroplanes. By 1918 more than 200 different types of aircraft had been developed by Britain alone.

The end of the war saw a collapse in the demand for new military planes, and the old ones could now be bought second-hand for next to nothing. Additionally, thousands of trained military pilots found themselves demobilised and seeking new outlets for their skills. Two of them, John Alcock and Arthur Brown, stunned the world in June 1919 by successfully flying a converted Vickers Vimy bomber clear across the Atlantic Ocean from Canada to Ireland and thereby winning a £10,000 prize from the *Daily Mail*. If aeroplanes could now cross oceans, mused the press, could they also carry fare-paying passengers?

They only had to wait two months for their answer. On 25 August, 1919, at 9:10 a.m. a converted de Havilland 4A bomber operated by Aircraft Transport and Travel (AT&T), took off from Hounslow Heath Aerodrome and flew to Le Bourget, Paris in a little under two hours.

*AT&T operated Britain's first scheduled passenger flight on 25 August 1919.*

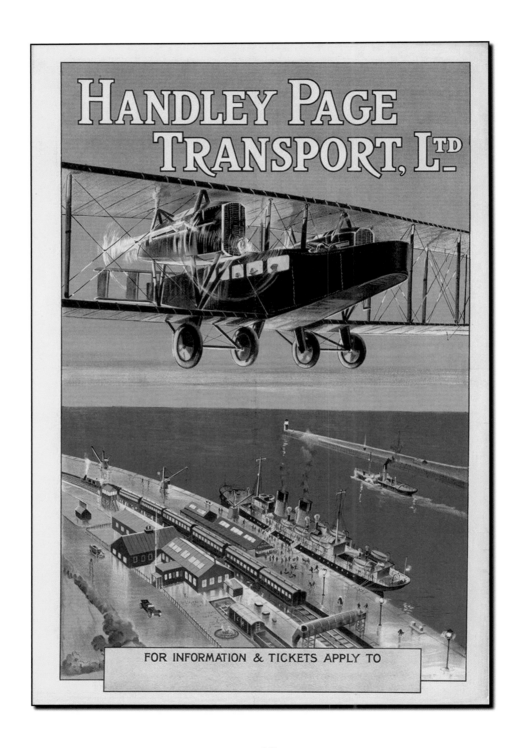

On board were Devonshire cream, newspapers, mail, and a single journalist from the *Evening Standard* who became the first ever passenger on a scheduled flight. A new industry had been invented; it would later be called the airline business.

Competitors quickly followed. AT&T was soon joined by Instone Air Line and Handley Page Transport offering services from London to Brussels and Paris. But the operating costs were enormous and, initially, only a tiny number of passengers were willing to either take the risk or pay the high fares of around £5 (roughly £250 in today's money). By October 1920, Handley Page had to stop its service, while six weeks later, AT&T cancelled its own scheduled flights. By February 1921, all British scheduled air services had ended due to the simple fact that the airlines were insolvent.

*A Vickers Vimy of Instone Airline.*

Like many others in the British Establishment, Winston Churchill, then Minister of War and Air, had not been an early advocate of commercial flying. He ignored an earlier recommendation from the Civil Aerial Transport Committee that the British government should help the post-war aircraft industry and in 1920 had said, 'Civil aviation must fly by itself.' He was, however, eventually persuaded to make some small subsidies, and these were enough to get the companies back in the air. By March 1921, Handley Page had reopened its London to Paris service. In the following year, government aid was again given to Handley Page, as well as Instone Air Line, Daimler Airway (who had absorbed AT&T), and the British Marine Air Navigation Company. The British airline business was back on its feet and between August 1919 and March 1924, the combined companies would carry nearly 35,000 passengers.

In an age where we take flying for granted, it is easy to forget how adventurous early airline passengers were thought to be. In the 1920s, the writer Virginia Woolf fled to the Continent with her lover, Vita Sackville-West. Their respective husbands, Leonard Woolf and Harold Nicholson, traced them to France and immediately flew there to bring them home (the return journey was by train). Virginia later wrote that what most touched her was not simply that her husband, Leonard, forgave her, but that he had been brave enough to fly to her in an aeroplane! Nonetheless, despite the perceived dangers of early air travel, it was actually safer than many thought.

*KLM was formed as early as 1919 and quickly developed the concept of a 'flag-carrying' national airline.*

During the first five years of British airline operations, only five passengers and six crew members were killed.

Although the British government might have been cool towards early commercial flying, other European countries took an entirely different view. In Holland, the government had formed their state-owned KLM as early as 1919 and quickly developed the concept of a 'flag-carrying' national airline whose purpose was, in part, to enhance Dutch prestige abroad. Other countries soon followed suit and by the early 1920s the British could see themselves being left behind. In 1923, a government committee was formed to review the policy of subsidising privately owned airlines. They quickly concluded that a single financially strong company was a lot better than four small, weak ones. In 1924, Handley Page Transport, The Instone Airline, The Daimler Airway and British Marine Air Navigation were all merged into a single airline to be known as Imperial Airways. Imperial would be a private company with private shareholders, but it would be subsidised to the tune of a million pounds over ten years and the government would appoint two of the directors.

*Waddon Aerodrome (later Croydon Aiport) circa 1922.*

# CHAPTER THREE

## Imperial Airways and the flying boat era

The clue was in the name. Unusually for a flag-carrying national airline, the word *British* was nowhere to be seen. In the 1920s, Britain's prosperity and standing in the world were still dependent upon a vast, far-flung, overseas Empire and the raison d'être of the new company would be to connect the mother country to the furthest corners of its realm. In 1924, when most aircraft had a range of only a few hundred miles, it was an ambitious step.

It did not start well. No sooner was the new airline formed than the pilots went on strike and all of the aircraft were grounded for three weeks. Even when the crews returned to work, the fleet that Imperial Airways had inherited from the four founding airlines was found to be poorly suited to the task: seven de Havilland DH 34s, three Handley Page W8Bs, two Supermarine Sea Eagle flying boats and an ancient Vickers Vimy. Most were obsolete and at least five were completely unserviceable. Worse still, the long-range aircraft

*An Armstrong Whitworth Argosy of Imperial Airways.*

# IMPERIAL AIRW

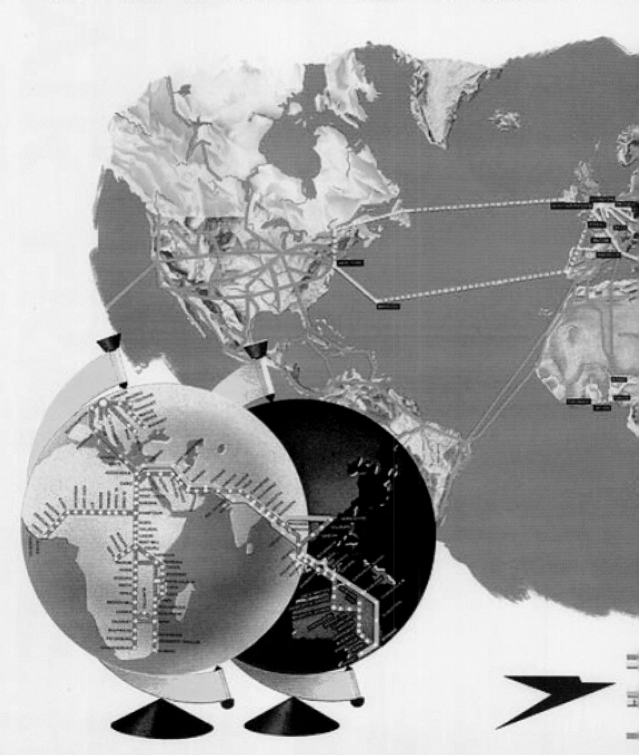

ENSIGN AIR LINER

EMPIRE FLYING-BOAT

...ES OPERATED BY IMPERIAL AIRWAYS & COMPANIES IN ASSOCIATION

...ES PROJECTED OR UNDER INVESTIGATION FOR OPERATION
...R IN CONJUNCTION WITH IMPERIAL AIRWAYS

...ROUTES OPERATED BY OTHER AIR TRANSPORT COMPANIES

*An Imperial Airways HP.42 over Croydon.*

*As the flying boats grew progressively larger and faster, so they became more sumptuous.*

Imperial needed to traverse the Empire were still to be designed, never mind built, and many of the routes they were to fly had not yet even been surveyed.

On 26 April 1924, a daily London to Paris service began with a DH34. A Southampton to Guernsey service commenced on 1 May, with London to Brussels, Ostend and Cologne following on 3 May, along with a summer service from London to Basle and Zürich via Paris. The new airline operated its London services from Croydon Airport (then called Waddon Aerodrome), conveniently located twelve miles south of the City and easily reached from the stockbroker belt.

*An Imperial Airways HP.42 at Sharjah.*

Paris quickly became a popular destination. Ernest Hemingway would later describe the 1920s city in his autobiographical work, *A Moveable Feast*. The young writer had lived there and mingled with most of the other major artists of the era: Pablo Picasso, James Joyce, Josephine Baker, F Scott Fitzgerald, Salvador Dali and many more. The French economy boomed from 1921 until the Great Depression reached Paris in 1931. This period, called *Les années folles*, saw Paris re-established as the capital of art, music, literature and cinema. Wealthy Britons flocked there and Imperial Airways, with its overtones of raffishness and perceived danger, became the chic way to arrive. The company responded with two innovations which continue to be imitated by other airlines to this day. When they received their first large aircraft, the three-engined Armstrong Whitworth Argosy, they installed a steward who would, 'Point out places of interest en route, attend to the comfort of passengers and serve light refreshments from the buffet.' The second novelty occurred when the Paris service became the world's first branded air route, having been dubbed *Silver Wing*. The return fare for the 2 hour 30 minute trip was £11.11s.0d, more then twice the average weekly wage. The name stuck and British European Airways (BEA) would continue to use it until 1974. Flying was no longer merely adventurous; it had become glamorous too. It would remain so until, and a little beyond, the advent of wide-body aircraft in the early 1970s.

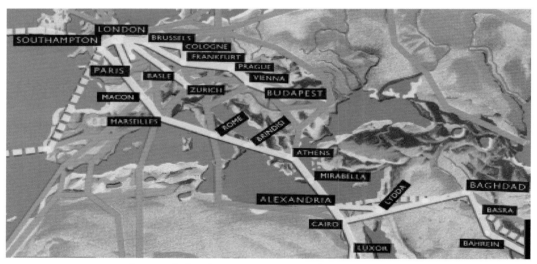

*Imperial Airways routes in 1937.*

Meanwhile, Imperial Airways had begun the awesome task of surveying thousands of miles of long-haul air routes across the British Empire. The new airline would have to fly very long sectors over

deserts, mountains and oceans in high temperatures and unpredictable weather – an infinitely greater challenge than their short-haul routes within Europe. Alexandria and Cairo became the principal hubs to connect passengers from London with flights to the Empire. The Cairo to Karachi route had been fully surveyed by 1925, but Cairo to Basra proved to be especially difficult. It involved an extremely long sector of completely featureless desert with little chance of the crew being able to identify anything on the ground. Incredibly, the problem was solved by the simple but hugely expensive means of ploughing a furrow, several hundred miles long, across the sand. It was probably the longest furrow in history!

*By 1928, Short Brothers had delivered the first of their three-engined Calcutta flying boats, which could carry fifteen passengers.*

Another challenge was that few cities in Africa or Asia then had airports. As the aircraft grew larger and needed more than a conveniently flat field, this became a taxing problem. Imperial had inherited two small Supermarine flying boats from British Marine Air Navigation, but they carried just six passengers and their limited range meant they could only be used on short hops to the Channel

Islands and France. Could a British aircraft manufacturer produce a much larger flying boat with the range and payload needed for Imperial's long-range routes?

In 1924, Short Brothers, Britain's oldest aircraft manufacturer, produced the first of a series of three flying boat designs known as the Singapore. In 1927, the Singapore I was made famous by Sir Alan Cobham, when he, his wife, and crew made a survey of Africa, and covered about 23,000 miles. By 1928, Short Brothers had delivered the first of their three-engined Calcutta flying boats, which could carry fifteen passengers and had a range of 650 miles. Imperial Airways used them to fly the Mediterranean to Karachi leg of the Britain to India route.

*Alan Cobham begins Imperial Airways' route survey flight from London to Cape Town.*

By 1931 Shorts had developed the S.17 Kent flying boat which had enough range to fly from Mirabella in Crete, to Alexandria in Egypt, without the need for refuelling stops. Its increased payload also meant that they could carry a steward and serve light inflight meals. Imperial Airways finally had aircraft that could properly meet the challenge of serving their far-flung Empire routes.

The Short Kent marked the beginning of another tradition which would be maintained by Imperial Airways and later BOAC. Each of the three aircraft was christened with its own unique name: *Scipio*, *Sylvanus* and *Satyrus*, and they were referred to within the company as the *Scipio Class* flying boats. The senior management of Imperial Airways would then have been classically educated, and so would most of the passengers. There was no embarrassment about using their aircraft to commemorate figures from Roman and Greek antiquity, even though their names would have meant little to the common man. As late as the 1970s, BOAC maintained the tradition of always having an aircraft in their fleet named *Canopus*. The last was a Super VC10 (registration number G-ASGD); her *Canopus* name plate was re-discovered when she was finally broken up by the RAF for spares in the 1990s.

In order to take off and land on water, the flying boats needed large, bulbous hulls, which meant they had considerably more internal space than the pencil-like fuselages of land planes. Imperial Airways put that additional space to good use – not only did flying boat passengers benefit from large seats with very generous leg room, but they could also enjoy lounges, sleeping berths, food freshly prepared in spacious galleys and even promenade decks. On a flying boat, which generally flew quite low, you could sit in a comfortable lounge while sipping your favourite cocktail and watching, from a panoramic window, the migrations of animals across the vast African plains below.

Travellers to the Empire would soon be enjoying a level of luxury previously only known aboard seagoing passenger liners, and this was an era in

*G-ASGD: the last aircraft named Canopus.*

*Replica of a Pan American Boeing 314, Foynes, Ireland.*

which the great liners were very luxurious indeed. In Evelyn Waugh's novel *Brideshead Revisited*, Charles Ryder and his wife decide to hold an impromptu party in their cabin aboard the Queen Elizabeth. When Celia Ryder asks the steward if he can organise some caviar, a large bowl of beluga arrives between the wings of a swan elaborately carved from a solid block of ice. Because Imperial Airways needed to lure passengers who were used to this kind of luxury at sea, no expense was spared trying to match it in the air. As the flying boats grew progressively larger and faster, so they became more sumptuous.

The steam ship lines, however, were not Imperial Airways only competitors. By 1927, an American entrepreneur named Juan Trippe had raised enough finance to begin running a flying boat service from Key West in Florida to Havana, Cuba. Trippe would become one of the airline industry's greatest innovators and his company, Pan American, was an unofficial 'flag carrier' for the

United States until its demise in 1991. Pan Am quickly expanded with routes throughout Central and South America using an ever-growing fleet of flying boats which he named Clippers. By 1933 they were ready to cross the Atlantic and on 5 July a Pan Am Clipper arrived in Foynes, Ireland at the same time that an Imperial Airways flying boat left for Canada. Not only could passengers now fly between the old world and the new, they even had a choice of carriers. For the next half-century Imperial Airways, followed by BOAC, would compete fiercely with Pan Am on the North Atlantic and elsewhere. Although Imperial Airways' primary purpose was always to connect Britain to its Empire in the south and east, the North Atlantic would become a hugely important route in terms of both prestige and profitability, and competition from Pan Am would constantly spur them to match or improve upon Trippe's many innovations.

If there was a moment in history when flying became truly glamorous, then this was surely it. One of the limitations of flying boats was that they could only operate during daylight hours as there was no safe method of taking off or landing on water at night. Every evening the flying boat would land in a

*Imperial Airways Short S.26 flying boat.*

river estuary or inland lake and the passengers would be taken to a hotel for the night – always the best hotel in town. A typical journey might be from Southampton to Marseille, and then onwards to Rome, Brindisi, Athens, Alexandria, Khartoum, Port Bell, Kisumu and finally by land-based aircraft to Nairobi, Mbeya and eventually Cape Town. Despite the many stops, it was still much faster than the traditional Union-Castle Line steamers, which could take more than two weeks to sail from Southampton to Cape Town.

Ross Stainton, a trainee manager in Cairo who later became the Chief Executive of BOAC, recalled how one of his duties was to visit not only Shepheard's Hotel (the leading hotel in Cairo and one of the most celebrated hotels in the world) but also the night spots and even the classier brothels, in order to make sure that every passenger was safely returned to the aircraft in time for its departure. If anyone was delayed, the obliging captain would normally wait until they could be found. It was an era in which the passengers were invariably too important to be left behind.

Flying boat passengers also needed a sense of adventure to meet the challenges of their destinations. Andy Carlisle, who was a flight

*The spacious galley on an Empire Class flying boat.*

engineer on a BOAC Short Solent, G-AHIR *Sark*, recorded this incident-packed few days after landing at Victoria Falls: 'I was on the wing after landing when we heard screams from one of the first passengers ashore. In the trees bordering the pathway up from the jetty, a lady passenger had come face to face with darkest Africa. A mighty python was draped along an overhanging branch and in its mouth was a half-swallowed monkey! Rest of night uneventful at Falls Hotel. Take-off next morning was at 06:30 and downstream, just about to unstick when nasty, grinding noise brought proceedings to a halt.

*BOAC Short Solent, G-AHIR 'Sark'.*

'We had hit a rock and not done a lot of good to the keel. I shinned down the central ladder and beheld a strange sight. The passengers were still sitting, belted in, while around them floated a miscellaneous assortment of newspapers, handbags, magazines etc.

'I got our steward, Bert Davies, bailing with a large, plastic waste bin while I kicked out one of the port windows and took on board the four inch suction hose from the service launch, tied up a couple of lines and slowly proceeded to a sandbar on the Northern Rhodesian bank and beached the waterlogged monster. Got the passengers clear, stripped off and did a bit of diving to try to find out by feel just how extensive the damage was. About a 12 foot (3.65 m) rent in the port keelson.

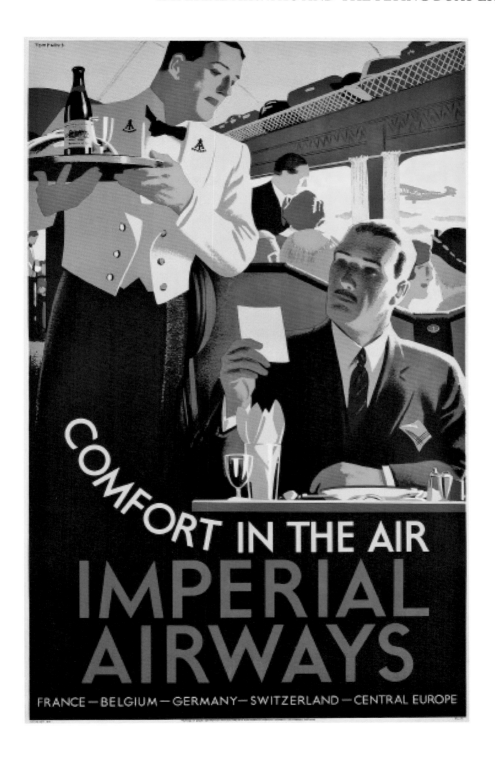

'We decided that as we were over that side of the Zambezi, we should night-stop in Livingstone for a first-light, minimum crew ferry flight. In the hotel where we were downing a few 'Tuskers', a tall, leathery individual approached and started giving us a hard time. It transpired that the sandbar where we had beached was the one that he used to stake out goats to catch crocodiles! And after two days of disporting ourselves almost naked in these same waters, now he tells us.'

On land too, air travel was becoming more pleasurable as the Handley Page company produced new aircraft for Imperial Airways that were significantly larger and more comfortable than their predecessors. Introduced in 1931, the HP.42 (for long-range eastern routes) and HP.45 (for European routes) were huge four-engined biplanes that were then the world's biggest airliners. They soon became the backbone of Imperial's land-based fleet. In an era where good airports with long runways were still rare, they had been designed to land and take off at low speed and from short grass strips. When the HP.42s were finally withdrawn from civilian service on 1

*HH Maharaja Umaid Singh of Jodhpur (second from right) with his brothers and sons at Croydon airport before their tour of Europe.*

September 1939, they had recorded almost a decade without any major accidents or fatalities. Like the flying boats, they carried a steward who served drinks and meals to passengers sitting in wide, deeply upholstered seats similar to those you would have found in a first class railway carriage.

These huge, lumbering biplanes were soon a familiar sight in the skies around Croydon and quickly became part of the mystique which surrounded Imperial Airways. The Maharaja of Jodhpur loved flying so much, that on his visits to London he regularly chartered Imperial Airways' HP.42s for cocktail parties in the sky above London! A passionate pilot himself (he commissioned Rajasthan's first airport), history remembers him for constructing the Umaid Bhawan Palace. Now a luxury hotel, it was the last great private palace to be built in India.

As the media regularly pointed out, perhaps this flying business was here to stay after all.

*A memorial to the flying boat era in the Victoria Falls Hotel.*

# CHAPTER FOUR

## The Speedbird

What today is known as 'corporate branding' barely existed in the 1930s, but Imperial Airways skilfully used all the resources of the then burgeoning advertising industry to create an image of exclusiveness and lavish comfort. Their posters, many produced by leading artists of the era, are still design classics and originals can be sold at auction for thousands. More importantly, Imperial Airways' logo would become the world's most recognised airline motif and the only one that most people could name: the Speedbird.

The effect their management were striving for, and largely achieved, was to create what modern marketeers call a 'luxury brand'. With air fares that represented more than a year's wages for many, and aircraft that could rarely carry more than a couple of dozen passengers, there was little point in trying to reach the average family. Imperial Airways' advertising was directed unashamedly at the tiny proportion of wealthy people who could actually afford such an expensive product.

*Charles Lindbergh – America's Lone Eagle – had flown a single-engined Ryan monoplane from New York to Paris in one hop.*

*Imperial Airways' posters, many produced by leading artists of the era, are still design classics and originals can be sold at auction for thousands.*

They were helped in this by a public relations coup which transformed public attitudes to flying in general and pilots in particular. In the years following World War One, Harry Houdini, Babe Ruth and Charlie Chaplin achieved extraordinary fame, but they would all be eclipsed in May 1927 when a hitherto unknown airmail pilot became the most famous man on earth in just 33 hours.

Charles Lindbergh – America's Lone Eagle – had flown a single-engined Ryan monoplane from New York to Paris in one hop, and to this day he is remembered as the first pilot to fly across the Atlantic. But in truth, he *wasn't* the first pilot to fly across the Atlantic – in fact, he wasn't even the second or the third. As early as 1919, a Curtiss seaplane had successfully crossed the Atlantic and just a month later the British aviators Alcock and Brown accomplished the extraordinary feat of flying across the Atlantic non-stop. Then an airship did the same thing and before long the South Atlantic had been traversed. Finally, two Portuguese pilots managed to fly across the Atlantic at night and in the wrong direction – against the prevailing winds!

Today, few of these pilots are remembered by anyone. For some years Alcock and Brown's flight was commemorated by a statue which overlooked the north runway of London's Heathrow Airport, but their monument was later moved to an obscure technical college (it's now in Ireland, close to the spot where they landed). Lindbergh's aircraft, however, named the Spirt of St Louis, is preserved at the Smithsonian Institution in Washington and is the first exhibit you see when you enter what is arguably the world's foremost aerospace museum.

So how did Lindbergh carve his name so effectively into history? Even today he is still considered one of the most famous Americans of all time. The answer, in case you hadn't guessed, is that Lindbergh had good PR on his side. In 1919, a New York hotelier named Raymond Orteig put up a $25,000 award for the first successful non-stop transatlantic flight, specifically between New York City and Paris. Over the next eight years numerous attempts were made, but all failed, and many of the pilots were killed. In 1927, Charles Lindbergh decided that he would try, and approached several aircraft manufacturers. He quickly discovered that his obscurity was a major disadvantage, the big plane makers would only provide an aircraft for a 'known' pilot and nobody knew Lindbergh. He responded by

*Alcock and Brown's statue.*

hiring a PR company and soon persuaded several St Louis businessmen, along with the Ryan Aeronautical company, to finance his flight and build him a specially designed machine.

Lindbergh landed at Le Bourget Aerodrome on the outskirts of Paris at 10:22 p.m. on Saturday, 21 May. He had found the airfield from the headlights of tens of thousands of spectators' cars caught in 'the largest traffic jam in Paris history'. A crowd estimated at 150,000 stormed the field, dragged Lindbergh out of the cockpit, and literally carried him around above their heads for nearly half an hour. Between 20 July and 23 October, 1927, Lindbergh visited 82 cities in all 48 states of the United States, and was seen by more than 30 million Americans, one quarter of the nation's population.

A 'Lindbergh boom' in aviation had begun. His flight was a bigger news event than the first landing on the moon would be decades later. The volume of mail moving by air increased by 50 per cent within six months, applications for pilots' licenses tripled, and the number of planes quadrupled. Lindy was now the most famous person in the world. Overnight, aviation had gone from an obscure niche to a major international industry and so had public relations. These lessons were not lost on the marketing men of the new airlines.

In 1932 Imperial Airways approached Theyre Lee-Elliott, one of London's leading graphic artists and painters, and asked him to create a symbol which they could easily reproduce on their posters and luggage labels. Lee-Elliott had been strongly influenced by the avant-garde work of Edward McKnight Kauffer, and his designs echoed Kauffer's angular bird forms in his 1918 poster for the *Daily Herald*. Other notable works by Lee-Elliott include posters for the London Underground and the Airmail logo. Many of his paintings and original artworks can still be seen at the Victoria & Albert Museum.

Lee-Elliott came up with a minimalist, simple, linear design representing a stylised bird in flight which soon became known as the Speedbird. In time it would become the best-known airline logo in the world and remains widely recognised, even to this day (at the time of going to press, British Airways has reintroduced it on one of their 747s). By 1938 it was appearing on the nose of aircraft, which was unusual in an era where most airlines, including Imperial Airways, kept their aircraft in a bare-metal finish with minimal signwriting. With the outbreak of war, BOAC aircraft had to be painted in wartime camouflage, but most still sported a Speedbird.

*Imperial Airways skilfully used all the resources of the burgeoning advertising industry to create an exclusive and luxurious image.*

*In the early 70s, Terminal 3 Heathrow was a sea of blue and gold.*

In the 1950s the Speedbird was prominently featured on the tail, initially in blue on a white background, and later in white on a dark blue background. As air traffic control developed, so 'Speedbird' became the call sign of BOAC flights and it is still the call sign of British Airways.

When Giles Guthrie was appointed Chief Executive in 1963, he decided to completely redesign the livery of BOAC aircraft, feeling that the 1950s scheme was now looking dated. He specified, however, that the Speedbird must remain on the tail. The result was a new, modern look with a gorgeous dark blue tail and cheatline, and a much larger Speedbird emblazoned on the tail in shining gold. It was an eye-catchingly beautiful design and in an era of strong airline logos it stood out at every airport that BOAC served. The new gold Speedbird, which the airline retained until the merger with BEA, helped define BOAC as a visible icon of the 1960s and it has rarely, if ever, been bettered by any airline.

In 2019, as part of their 100th birthday celebrations (remember AT&T in 1919), British Airways painted a Boeing 747-400, registered

G-BYGC, in the same magnificent colour scheme; it will remain in this livery for the rest of its service with BA. As the Chairman and CEO of British Airways, Alex Cruz, put it: 'So many British Airways customers and colleagues have fond memories of our previous liveries, regularly sharing their photos from across the globe, so it's incredibly exciting to be re-introducing this classic BOAC design. Our history has shaped who we are today, so our centenary is the perfect moment to revisit our heritage and the UK's aviation landscape through this iconic livery.'

*Britsh Airways' Boeing 747-400 G-BYGC has been repainted in BOAC livery.*

Contemporary Art Moderne style would also appear in Imperial Airways imposing new London headquarters. In 1937, construction started on the Empire Terminal in Buckingham Palace Road. It was designed by the architect Albert Lakeman and included an elaborate statue over the entrance by the sculptor Eric Broadbent: *Speed Wings Over the World.* The terminal had its own platforms for train connections to Imperial's flying boats at Southampton and coaches to its landplane base at Croydon Airport. With the possible exception of the Pan Am Building in Manhattan (1960–63), it was probably the grandest city-centre building to be erected by any airline in the world. It was still offering British Airways' passengers a London check-in facility as recently as the 1980s, but today the Grade 2 listed building is home to the UK's National Audit Office.

*In 1937, construction started on the Empire Terminal in Buckingham Palace Road. It was designed by the architect Albert Lakeman.*

## CHAPTER FIVE

## The birth of BOAC

The British Overseas Airways Corporation was a child of the war. Had there not been a second global conflict only a generation after the first, the company would not have existed. In the post-war years BOAC would firmly establish itself as an iconic global airline but, quite apart from the powerful symbols it created in the minds of the public, there were also two unique feats in its history, one of which was widely publicised and the other hardly at all. The first was the introduction of the world's first jet airliner in 1952 and the second was the company's extraordinary and largely unsung achievements during the Second World War.

From 1940, most of Continental Europe was either occupied, or otherwise manipulated, by Nazi Germany. To stand any hope of winning, the British had to maintain transport links with their overseas Empire and the rest of the world. For BOAC, newly nationalised and tasked with supporting the war effort, the challenge was enormous.

*Only the Short Empire flying boats were really up to the task of flying the lengthy and hazardous Horseshoe Route. 'Cleopatra' is seen over Durban.*

*The film star Leslie Howard was killed in 1943 when the Germans shot down a BOAC-KLM DC3.*

Connecting Britain to her far-flung Empire had been testing before the war but it now became infinitely more difficult and dangerous. BOAC's vital services to Africa, Asia and Australia were staged through Egypt (Alexandria for flying boats and Cairo for land-based aircraft), but how could you reach Egypt without flying over occupied territory? The answer came in the shape of an ancient symbol of luck – the Horseshoe Route.

BOAC aircraft flew a wide route over the Bay of Biscay to Lisbon in neutral Portugal, avoiding France and Spain, and from there to

Gibraltar, then onwards to Malta and finally to Cairo. The route involved enormous risks, aircraft were shot down and lives were lost
(including that of the film star Leslie Howard), but the service was
maintained. When Malta came under attack, they switched to the
longer West Africa route: Lisbon, Bathurst, Freetown, Lagos and
then by landplane to Khartoum. Much of this flying had to be done
with obsolete, unsuitable aircraft including the Armstrong Whitworth Ensign and de Havilland Albatross. Only the Short Empire
flying boats were really up to the task and just a handful of these

*Lockheed Lodestar, G-AGCM 'Lake Marut', of BOAC, flying over Cai-*
*ro before landing at Heliopolis.*

had long-range tanks but, by and large, BOAC managed to get
through. Had they not done so, the results for Britain would have
been catastrophic.

In 1939, the RAF had virtually no transport capability of its own;
its fleet was comprised largely of fighters, bombers, training aircraft
and some flying boats. Neither could it spare many of its pilots, having begun a massive recruitment and training programme for the
aircrew it knew it would soon need for operations. The government
was better prepared than in 1914 and understood that it would
shortly have to move huge quantities of personnel and material over

great distances where surface transport would be too slow or impractical.

On 3 September 1939, the day war broke out, all civilian airlines were placed under the direct control of the government and all unauthorised civilian flying was halted. The two leading British airlines, Imperial Airways and British Airways Ltd, were to be combined into a single organisation whose sole purpose was to support the war effort.

British Airways Ltd was a quite different airline to Imperial. It had been created in 1935 by the merger of several smaller airlines and provided domestic services within the United Kingdom as well as short-haul flights to a number of destinations in Europe. Unusually for the day, and unlike Imperial Airways, it had been allowed to buy both German and American aircraft. In 1937 it had acquired seven Lockheed Electras from the United States and these fast, modern machines would prove to be invaluable during the war. In 1938, one of these Lockheeds had flown Neville Chamberlain back from Munich to Heston Aerodrome where he gave his infamous 'Peace in

*In 1938, a British Airways Lockheed Electra had flown Neville Chamberlain back from Munich to Heston Aerodrome.*

our times' speech, having negotiated what he thought was a reliable accord with Hitler. The newsreel of that event must be one of the most widely shown airport arrivals in history.

Before the war, Imperial Airways had been ably managed by George Woods Humphery. He, however, had fallen out with the government and had resigned to be replaced, a few weeks before war broke out, by a political appointee, Sir John Reith, the former Director-General of the BBC. Reith staunchly believed that Imperial Airways, like the BBC, should be run purely as a public service and not for profit. This meant a significant change because both Imperial Airways and British Airways had previously been private companies with largely private shareholders. Although they had been subsidised and regulated by the government, they had nonetheless retained a strong element of independence.

Reith won the day and convinced the government that all private shareholders should be bought out and the new company, which he named BOAC, would be a wholly state-owned corporation which would have a public service ethos and not be run as a commercial business. This crucial transformation would shape (and to some degree dog) BOAC right up until the merger with BEA in 1974. In

*Sir John Reith, the former Director-General of the BBC, became Chairman of Imperial Airways shortly before it was merged into BOAC.*

the post-war years, a political debate would rumble on over whether major airlines should be publicly or privately owned. Not until 1987 would the new British Airways plc be privatised and shares again sold to the public. Reith's tenure was short, he left BOAC in 1940 to become Minister of Information in Chamberlain's government, but his influence was huge and controversial.

France had not yet been occupied by the Germans, and its rapid fall in 1940 had not been anticipated. A 'Bomb Line' had been drawn across England to mark what was believed to be the maximum range of German bombers. Consequently, BOAC's operations were to be based to the west of it. Flying boats would be operated from Pembroke Dock, Falmouth, and Poole while land-based planes would fly from Whitchurch near Bristol and Exeter. With obsolete aircraft, a serious shortage of spares and civilian crews, BOAC quietly got on with their unenviable task. Once France had fallen in the summer of 1940, all of BOAC's UK bases were comfortably within the range of German bombers and some were attacked.

When the Germans occupied Holland, a number of KLM aircraft that had been on British soil were interned. The Dutch government-in-exile agreed that they could be used by BOAC and the British government, after some reluctance, eventually agreed that they could be flown by Dutch crews who would be based at Whitchurch. BOAC used them to replace their de Havilland Albatross aircraft on the Bristol to Lisbon section of the Horseshoe Route and operated them with BOAC flight numbers. The aircraft were repainted in camouflage, with British civil markings and the same red, white and blue stripes that had been applied to other BOAC aircraft. The only thing they lacked was the Union Flag, although they were later marked with Dutch bird names under the cockpit windows. The interiors remained in KLM colours and markings.

Portela Airport, seven kilometres from Lisbon, was also used by Lufthansa, the German civilian airline. Consequently, it soon became a hotbed of spies with British, German, Soviet, and American agents all closely watching the traffic, and each other. They were especially interested in the Lisbon to Whitchurch route - it frequently carried agents and escaped prisoners of war to Britain. German spies in the terminal building meticulously recorded every passenger who arrived or departed. Harry Pusey, BOAC's operations officer

*Portela Airport in neutral Portugal in 1943.*

in Lisbon between 1943 and 1944 later said that the area was, 'Like the film *Casablanca*, but twentyfold.'

In the early part of the war the Germans respected Portuguese neutrality and did not attack British aircraft crossing the Bay of Biscay on their way to and from Lisbon. Things began to change, however, in 1942, as the German navy began to use the French Atlantic ports to attack British shipping. On 15 November the KLM DC3 *Ibis*, which had been re-registered G-AGBB by BOAC, was attacked by a Messerschmitt 110 and, although seriously damaged, managed to return to Lisbon where she was repaired.

Early in 1943 the same aircraft, *Ibis*, was attacked by six Bf 109 fighters. Incredibly, the Dutch pilot managed to first dive to sea level, then pull up into cloud and, once again, the aircraft made it back to Lisbon where she received extensive repairs. A few weeks later she would be less fortunate. On 1 June 1943 *Ibis* was carrying four crew and thirteen passengers from Lisbon to Whitchurch, including the British film star, Leslie Howard. Two hundred miles north of Spain they were attacked by eight German Junker JU 88 fighters, but this time they were unable to escape. The aircraft crashed into the Bay of Biscay and there were no survivors.

Controversy surrounds the tragedy to this day. One theory is that the Germans targeted the DC3 because they believed Winston Churchill was on board. Another suggests that the Germans knew Leslie Howard was a passenger and they believed him to be a British spy. Interviews conducted with the German pilots after the war suggest a simpler explanation: the JU 88s had been ordered to find and escort two German U Boats and they had come across the DC3 by chance. In hazy visibility, they recognised it as an enemy aircraft but did not realise it was a civilian airliner.

Nor were the Germans the only danger. On 15 February 1942 a BOAC B-24 Liberator with the registration G-AGDR was returning to Bournemouth, Hurn Airport from Cairo. While over the English Channel it was shot down by an RAF Spitfire whose pilot had misidentified the Liberator as a German Focke-Wulf Condor. The BOAC plane crashed into the sea close to the Eddystone Lighthouse, and all nine on board were killed. Ironically, a Danish aircraft interned

*BOAC B24 Liberator landing at Prestwick.*

by the British in 1940 and subsequently flown by the RAF was also a Focke-Wulf 200 Condor! It was, perhaps, fortuitous that the RAF wrote the German plane off in a landing accident before one of their own fighters had a chance to shoot it down. Following this tragedy, the RAF significantly ramped up aircraft recognition training among its pilots.

Other BOAC crews would die in accidents, many of them exacerbated by obsolete or unserviceable aircraft, bad weather, fuel exhaustion (BOAC had to push the ranges of their aircraft to the absolute limit) and the difficulties of flying over territory that was either occupied by the enemy or completely blacked out.

The supply of suitable, modern aircraft was a continual problem. In 1939 Pan American had begun to operate the gigantic Boeing 314 flying boat on their Pacific services. Big, even by today's standards, it had a wingspan of 152 feet (the wingspan of a modern 747 is 224 feet) and could carry up to 74 passengers at a cruising speed of 188 miles per hour with enough range to fly from San Francisco to Honolulu. Never slow to exploit a commercial opportunity, Juan Trippe used this new giant to offer passengers unprecedented levels of comfort. Pan Am's 314s had a sumptuous lounge and dining area, and the galleys were crewed by chefs from leading hotels. Men and women were provided with separate dressing rooms, and white-coated stewards served five

*BOAC Boeing 314 flying boat 'Berwick'.*

and six-course meals from silver service. It was a level of luxury that has rarely since been matched in the air.

When Pan Am reduced their order for 314s because the occupation of France had made it difficult for them to fly the Atlantic, they put three of the aircraft up for sale at a knock-down price. BOAC desperately needed a longer-range flying boat with a good payload so that it could continue to operate vital services to the USA and Canada, but at the height of the war it was almost impossible to persuade the Treasury to release precious US dollars for the purchase. In the end, Harold Balfour, a junior minister, went out on a limb and approved the purchase without the consent of the Cabinet. Churchill was furious and a full-blown political row followed, but the order was not cancelled, and BOAC received the three aircraft.

Churchill, who had never been very keen on subsidising civil aviation, later became a convert to these extraordinary planes. In 1942 he had sailed to America on the battleship *Prince of Wales*

to lobby Roosevelt for American support. The original plan was to return by sea via Bermuda but, while in Baltimore, he had gone aboard one of BOAC's new Boeing 314s, *Berwick*, and met her captain, John Kelly-Rogers. Churchill was impressed, and casually asked Kelly-Rogers if *Berwick* was capable of

*Churchill asked if he could take the controls.*

flying him to Bermuda and then home. When the captain said yes, Churchill immediately changed his plans and he, along with his considerable entourage, made the trip by air. *Berwick* had to be specially lightened for the journey which was at the limit of her range. Accompanying Churchill were, among others, the First Sea Lord, Admiral Sur Dudley Pound, Air Chief Marshall Sir Charles Portal and Lord Beaverbrook, Minister for Aircraft Production.

During the flight, Churchill, who had previously had a few flying lessons, asked if he could take the controls and was photographed flying the huge aircraft with a cigar clenched between his teeth. The trip to Bermuda was uneventful but during the subsequent flight from Bermuda to Plymouth, the navigator underestimated the wind and the aircraft came close to the French coastal town of Brest, a heavily defended German naval base. A Luftwaffe squadron was scrambled to intercept the plane, but fortunately failed to find it. *Berwick* flew for the last two hours of the flight with radio silence and

approached England from an unexpectedly southerly direction. Churchill later recorded: 'Six Hurricanes from Fighter Command were ordered to shoot us down but failed in their mission. I thought perhaps I had done a rash thing that there were too many eggs in one basket. I had always regarded an Atlantic flight with awe. But the die was cast. Still, I must admit that if at breakfast, or even before luncheon, they had come to me to report that the weather had changed and we must go by sea, I should have easily reconciled myself to a voyage in the splendid ship which had come all this way to fetch us.'

When Churchill finally debarked at Plymouth, Kelly-Rogers said to him: 'I never felt so much relieved in my life as when I landed you safely in harbour.'

CHAPTER SIX

## The DH-98 Mosquito

During its relatively short history (just 35 years) BOAC operated no less than 43 different types of aircraft. When formed in 1939, they had inherited the combined fleets of Imperial Airways and British Airways Ltd. The result was a mixed bag of flying boats, some modern American aircraft including the Lockheed Electra, and a significant number of obsolete airliners from an earlier era. These included such relics as the Armstrong Whitworth Atlanta, a ponderous four-engined design which had first flown in 1932 and carried only nine passengers at 115 mph! Thirty-five years later BOAC pilots would be test-flying Concorde at supersonic speeds.

Throughout the Second World War any number of military aircraft were 'civilianised' and transferred to BOAC, although 'civilianisation' rarely meant more than removing the guns and turrets.

*The Mosquito was based on the DH-88 Comet, which won the 1934 London to Melbourne air race.*

Many of them were poorly suited to the tasks of flying VIPs, troops and material over long distances and into parts of the world where there were few facilities and even fewer spare parts. The most remarkable, however, must surely have been the de Havilland DH-98 Mosquito. In 1943 it was one of the fastest and probably the most versatile aircraft in RAF service.

The original idea for the Mosquito had come from de Havilland themselves as early as 1938. They based the design on their DH-88 Comet, which won the 1934 London to Melbourne air race and is still considered to be one of the most beautiful aircraft of all time. (The original DH-88 survives to this day and is regularly flown by the Shuttleworth Collection.) De Havilland's concept was simple: build a light aircraft with two Rolls-Royce Merlin engines whose main defence would be its speed and manoeuvrability. In order to reduce weight, it would be constructed largely from wooden laminates – a material which de Havilland had extensive experience of.

The aircraft's phenomenal speed and agility meant that Mosquitoes were rarely shot down by the Luftwaffe. In fact, so valuable was the Mosquito that the Air Ministry would have needed a very good

*The original DH-88 survives to this day and is regularly flown by the Shuttleworth Collection.*

reason indeed to divert precious airframes away from the military and give them to what was ostensibly still a civilian airline.

The rationale was clear: one of BOAC's most vital tasks throughout the war was to maintain air services to neutral countries, and among the most important of those was the route to Sweden. Before the war, British Airways Ltd had already been operating a weekly service to Stockholm. When hostilities broke out and Norway was occupied, this was changed to a new route via Perth and Helsinki using Lockheed 14s. The direct *Stockholmsrute*, as the Scandinavians called it, was originally set up by the Norwegian government, exiled in London, to rescue Norwegians who had escaped from Nazi-occupied Norway and bring them to Britain. Several types of aircraft were used, but the route was mostly flown with Lockheed Lodestars operating from Leuchars in Scotland. As well as aiding escaping Norwegians, Sweden produced high-quality ball bearings which were vital to the war effort, and the route was quickly dubbed the 'Ball Bearing Run'.

*A BOAC Mosquito on its way to neutral Sweden.*

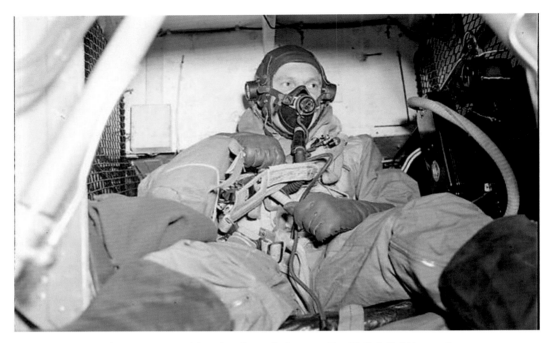

*A 'passenger' in the bomb bay of a BOAC Mosquito.*

The civilian Lockheeds were soon replaced with ex-RAF Hudson aircraft, but they were vulnerable to German fighters and when the Swedes had two of their own DC3s shot down during flights to Britain, minds were concentrated on finding a faster alternative. Some interesting conversations must have taken place at the Air Ministry and elsewhere. The Mosquito was arguably the most valuable aircraft that the RAF then possessed, and its complex, bonded wooden structure meant that they were not easy or quick to build. On the other hand, the government considered that maintaining the air route to neutral Sweden was vital and it was that argument which finally won the day.

The de Havilland Mosquitoes, painted in BOAC colours and flown by pilots in BOAC uniforms, began flying to Stockholm in 1943 during the short summer nights and periods of the Aurora Borealis. Their crews must have seen some spectacular views which would probably have been cold comfort given the risks they were taking. The specially lightened aircraft operated at very high altitude, over 35,000 feet, which meant they could overfly German-occupied

Norway where they were too fast and too high to be caught by conventional German night fighters. The exhaust shrouds had been removed from the engines, adding around 30 miles per hour to the Mosquitoes' already considerable speed. The Germans responded by fitting some of their JU 88-night fighters with GM-1 nitrous-oxide injection, which significantly increased their speed, but they never succeeded in shooting down a BOAC Mosquito.

The Mosquitoes carried mail and diplomatic baggage, as well as gold ingots and sovereigns which were used to pay for the ball bearings. Incredibly, they could also carry a single passenger in the aircraft's bomb bay. The passenger, once airborne, was completely unreachable by the crew of two. They would simply be given a parachute, an oxygen mask, a flask of coffee and some blankets to protect them from the cold!

One of the most important passengers was the notable Danish scientist Niels Bohr who was an expert in atomic research. When Bohr learned that the Germans planned to arrest him, he fled to Sweden from where a BOAC Mosquito brought him to Britain. During the flight, Bohr took off his uncomfortably small flying helmet and consequently failed to hear the pilot's instruction to turn on his oxygen supply as the aircraft climbed to high altitude over Norway. He lost consciousness from oxygen starvation and only woke up when the aircraft descended to lower altitude over the North Sea. The pilot, having heard no sound from the bomb bay for some time, had guessed what had happened and saved Bohr's life by descending.

*One of the most important passengers to travel in a BOAC Mosquito was the notable Danish scientist, Niels Bohr.*

In what must have been the most unusual service ever undertaken by BOAC, a total of thirteen Mosquitoes were operated, of which five crashed (none from enemy action) with several pilots and radio operators losing their lives. Between 1939 and 1945, 6,000 passengers and 500,000 tons of freight were transported between Stockholm and Great Britain.

Night maintenance work on one of the four Ghost engines which power the de Havilland Comet jet airliner. The Ghost has a basic advantage of efficiency through the direct entry of the air into the eye of the compressor—as is clear in this picture—and offers unprecedented standards in simplicity, robustness, accessibility and protection from fire and ice.

DE HAVILLAND JET ENGINES
for efficiency at speed and altitude

*The de Havilland Comet I changed commercial flying forever but had to be grounded after only two years.*

# CHAPTER SEVEN

## After the war

BOAC had a good war. The company had been created to perform what was arguably the most difficult and dangerous challenge ever presented to any civilian airline, but they got the job done while receiving little public recognition for it. Throughout the war, British propaganda had concentrated on the military operations of the armed forces and the stoic resistance of ordinary civilians on the home front. Few people knew that while the air battle was being fought in the skies above Britain, in full view of everyone, BOAC flying boats were crossing the Atlantic, Liberators were returning ferry pilots to Canada and civilian Mosquitoes were carrying nuclear scientists to Britain in their bomb bays.

In the 1945 General Election, the Labour Party under Clement Atlee won a surprising landslide victory and, in a time of austerity, ushered in a policy of nationalising many key industries. About 20 per cent of the entire economy including coal, railways, road transport, the Bank of England, civil aviation, electricity, gas, and steel were taken into public ownership. BOAC would remain a wholly state-owned corporation with no possibility of private shareholders. More importantly, two new wholly state-owned airlines would be created: British South American Airways (BSAA), which would fly to the Caribbean and South America and British European Airways (BEA), which would fly routes within Europe.

BOAC did not raise strong objections, possibly because they thought this division of routes would make it easier for them to raise the capital they needed for new aircraft and facilities (a consortium of shipping lines had already agreed to finance BSAA). Another reason may have been that BOAC's management, many of whom had come from Imperial Airways, still saw the Empire routes and the North Atlantic as their natural territory, rather than Europe or Latin America.

BSAA was short lived. In line with Brabazon Committee recommendations, they had been persuaded to buy a fleet of new Avro Tudor aircraft. But the Tudor, a development of the wartime Lancaster bomber, turned out to be unreliable and there was a string of fatal accidents, some of which remain unexplained to this day. By 1949 the Tudors had been grounded and the government ordered BSAA to be merged back into BOAC. BEA would stand alone until its merger with BOAC in 1974, but the existence of a separate European airline allowed BOAC to hone its image as a purely long-haul carrier and, in a post-war Britain where few people travelled outside Europe, they were able to retain some of the exclusivity and prestige which had always surrounded Imperial Airways.

When peace returned, BOAC inherited a seasoned management team who had met every conceivable difficulty the airline business could throw at them, coupled with the real risk of their aircraft being shot down by the enemy. They also inherited a mix-and-match fleet of obsolete airliners and more recent military aircraft that

*BOAC's Stratocruisers had a double-decked fuselage which lent itself to sleeping berths and even a downstairs cocktail bar.*

*BOAC was finally allowed to purchase five Lockheed Constellations.*

were hopelessly unsuited to the task of flying civilian passengers in peacetime. An immediate debate began which would continue throughout the 35 years of BOAC's existence, and beyond: should BOAC be allowed to use scarce US dollars to buy modern American airliners, or should they be obliged to stick with British-built machines? Pan American, BOAC's principal competitor on their most lucrative routes, had already placed an order for twenty Boeing 377 Stratocruisers (it was then the largest order in the history of the commercial aircraft business), and Lockheed was offering BOAC's competitors their fast, pressurised L-049 Constellation. In the meantime, BOAC was having to make do with aircraft such as the Avro Lancastrian, a civilianised version of the Lancaster bomber which carried a handful of passengers in a noisy, unpressurised bomb bay.

After the First World War a number of British aircraft manufacturers, including the legendary Sopwith, had been bankrupted when the demand for military machines abruptly dried up. The British government, anxious to avoid another collapse in what they now saw as a key industry, had formed the Brabazon Committee as early as 1942 to investigate the peacetime needs of the civilian airliner market. At that time, the British were still in the mindset that connecting Britain to its Empire was crucial. Few foresaw that the Empire would last barely twenty years following peace, and fewer still guessed that post-war aviation would be less about luxury and more about economy, preferring to stick with the pre-war notion that flying was for the rich and powerful rather than the common man.

Of the seven designs which the Brabazon Committee recommended, only two were commercially successful: the de Havilland Dove and the Vickers Viscount, and neither of those came directly from Brabazon's recommendations. The Dove had been privately developed by de Havilland and, thanks to Vickers' doggedness, the hugely successful Viscount was a much larger and better aircraft than the one envisaged by the Committee. Sadly, neither of these designs had enough range to be much use to BOAC (although small numbers of both would later appear in their fleet).

As for the designs that were intended for BOAC, the enormous Bristol Brabazon, which had been closely tailored to what Bristol thought were BOAC's needs, proved to be too expensive to operate profitably and none were sold. The Britannia was a good aircraft,

*The Britannia was a good aircraft but technical problems seriously delayed her entry into service.*

but delays caused by the unreliability of her engines and electrical systems meant she came into service too late to be profitable. As for the Comet I, which we shall return to later, she changed commercial flying forever but, tragically, had to be grounded after only two years following several accidents. By the time the much improved Comet IV came into service she, like the Britannia, was too late, having been made obsolete by the Boeing 707 and Douglas DC8.

*By the time the much improved Comet IV came into service she, like the Britannia, was too late, having been made obsolete by the Boeing 707 and Douglas DC8.*

The mistakes made by the Brabazon Committee cost BOAC dear but crippled the British aircraft industry. No complete civil airliners have been made in Britain since 2002 and of the many designs produced since the war, only a handful were commercially successful. Among the few winners were the de Havilland Dove, Vickers Viscount, BAC 1-11, HS 748 and BAe 146, but none of those suited BOAC's routes (the list of commercial failures is too long to reproduce here). To make matters worse, the British-built aircraft that did enter service with BOAC often had poor service records. Former BOAC Chairman, Sir Miles Thomas, would later say of the Handley Page Hermes, 'It had every possible fault except that it didn't blow up!' The Avro Tudor, initially operated by British South American

Airways but later by BOAC, was banished to cargo-only flying after a spate of accidents.

One post-war aircraft that did serve BOAC well was the Canadian built Argonaut. Canadair had cleverly adapted the Douglas DC4 airframe by adding Rolls-Royce Merlin engines which, along with other modifications including pressurisation, produced a fast, reliable aircraft that, although rather noisy for passengers, gave good service. Additionally, the fact that she had British engines and was built in Canada reduced her cost in US dollars, which pleased the British Treasury.

*The Canadair Argonaut served BOAC well.*

But neither the Argonaut, nor any of the other aircraft in BOAC's immediate post-war fleet, were suitable for their crucial North Atlantic routes. Following a lengthy argument, which would be repeated every time BOAC tried to buy American aircraft, the corporation was finally allowed to purchase five Lockheed L-049 Constellations and, a few years later, six Boeing 377 Stratocruisers. Parliament, the media, British aircraft manufacturers and the trade unions all loudly complained, but BOAC could now offer a North Atlantic service which rivalled that of Pan American.

Juan Trippe, however, always the innovator, was offering Pan Am's passengers more than merely modern aircraft. They could now, rather like the railway passengers of the past, have a choice of different classes of travel: first or economy. This was a revolutionary development which was alien to Imperial Airways' culture of charging all passengers high fares in exchange for luxurious travel with meals, drinks, hotels and ground transport etc. included. BOAC had to follow suit but did so with an innovation of their own: first class travel across the Atlantic would now be branded as 'Monarch Class' and marketed separately. BOAC's Stratocruisers had a double-decked fuselage which lent itself to sleeping berths and even a downstairs cocktail bar. During the long flight to New York, passengers enjoyed seven-course dinners and breakfast in bed; while female passengers received Speedbird beauty kits and everyone got an overnight bag. BOAC managers privately called it the, 'De-luxe, dollar-earning service'. Henceforth, on the North Atlantic routes, some Stratocruisers would be operated in all Monarch Class configuration, but economy class would be offered on other flights.

The quality of food also reached a peak. Pan Am had commissioned Maxim's-de-Paris, at the time the world's most famous restaurant, to produce the meals for their first class 'President' service. All the food was prepared in the restaurant's Paris kitchens, flash-frozen and delivered to staging points across the world. BOAC had no option

*A Class of its Own* by Chris French, GAvA.

but to offer meals on their Monarch service which were equally good. Although sumptuous flying boat standards would be retained for a few wealthy passengers, the democratisation of air travel had begun. Airlines were beginning to learn that reducing service and fares, while simultaneously increasing the number of seats, was good for the balance sheet. By 1950 all the destinations that BOAC served had serviceable airports and could now be reached by land-based aircraft. As a result, flying boat operations, which were expensive and did not lend themselves to economy class, were ended and the era of slow but luxurious flying with multiple stopovers

en route, came to an end. Air travel would now be faster and cheaper, but a little less romantic.

Before long, however, Juan Trippe's Pan American would launch another innovation which, like economy class, would change the airline business for ever. Trippe had long understood the importance of uniforms, he had already taken his pilots out of the military-style khaki, with belts and epaulettes, that most other airlines used and replaced them with navy blue blazers sporting gold buttons, gold rings on the sleeve, and white-topped caps. His reasoning was simple; he was competing with the shipping lines, so why not make Pan Am crews look like merchant marine officers on a smart ocean liner? Before long, practically every other airline, including BOAC, had followed suit.

Pan American flights now began to feature stewardesses in chic, tailored uniforms who had to meet strict rules on height, weight etc. and were clearly chosen, at least in part, for their appearance. If the girls got engaged, married or pregnant, they had to resign immediately. Before the war, all Imperial Airways crew, whether on the flight deck or in the cabin, had been exclusively men, and many of the stewards had been recruited from the shipping lines. BOAC did not immediately react, but when passengers started showing a clear preference for female cabin crew, they had little option but to relent. Initially, just one stewardess was carried on a Stratocruiser, but their popularity was such that BOAC quickly began recruiting women in significant numbers.

Airline marketing was soon featuring pretty girls in figure-hugging uniforms, and by the 1960s BOAC would be running blatantly sexual advertising. One advert, for the VC10, featured a steward-

All over the world B.O.A.C. takes good care of you

ess's pretty legs in an extremely short skirt with the strapline: 'Should BOAC stewardesses wear the mini skirt?' It was clearly a rhetorical question. In 1967 BOAC introduced short, translucent paper dresses which were worn by stewardesses operating from New York to the Caribbean. At the end of the decade BOAC turned to Clive of London, a designer who was then at the epicentre of swinging sixties fashion and commissioned a Terylene and cotton mini skirt in a choice of pink or blue. Despite the hemline already being well above the knee, many stewardesses shortened the skirts even further. Although this was technically against the uniform regulations, management generally turned a blind eye. Flying had always been glamorous, but from this point on it would be sexy too.

BOAC cabin crew were not allowed to accept gifts from passengers, but there were always exceptions. 'I began dating a sheikh in Bahrain,' recalled stewardess Katie Howe. 'He asked me what my dream car was, and I jokingly said, "A shocking pink Jaguar E-type!" He asked me to come to a Jaguar garage and, to my amazement, he presented me with a pink Jaguar E-Type. I was 22.'

# CHAPTER EIGHT

## The pilots

From the earliest days of commercial flying, safety had been a concern. While airline transport had never been quite as dangerous as many believed, aircraft accidents always drew lurid headlines and fear of flying was undoubtedly a deterrent in the minds of the public. BOAC's marketing people knew this was a hurdle they had to clear and soon discovered that cold statistics had little effect.

Often referred to as the father of public relations, Edward Bernays had published his seminal book, *Propaganda*, in 1928, and he is credited with having created modern public relations and lobbying techniques. As he put it: 'The conscious and intelligent manipulation of the organised habits and opinions of the masses is an important element in democratic society. Those who manipulate this unseen mechanism of society constitute an invisible government which is the true ruling power of our country. We are governed, our minds are moulded, our tastes formed, and our ideas suggested, largely by men we have never heard of... It is they who pull the wires that control the public mind.' The gist of

*The cockpit of a BOAC Boeing Stratocruiser.*

his philosophy is that if you want to change people's minds about something, then you need to inject emotion rather than mere facts into your messages.

Post-war aircraft had almost no automated systems. They were flown by human pilots, engineers and navigators, and (unlike modern aircraft) their cockpit controls were linked directly to the aircraft's control surfaces, engines and other systems without the intervention of computers. If they had an autopilot, it would rarely be more than a simple 'wing-leveller', in other words, a device that allowed the pilots to fly 'hands-off' but took no other decisions on their behalf. The safety of the passengers resided almost entirely in the training, experience and skill of the human beings who sat in the cockpit. BOAC had some highly experienced pilots and had always taken crew training very seriously. The marketing people came up with an idea which Bernays would have been proud of. Rather than highlighting the aircraft, whose technicalities were only of interest to a few and beyond the comprehension of most, would it not be better to put the *pilots* on centre stage and encourage the public to put

*When Captain Oscar Phillip Jones (better known as O P) retired in 1959, his career was featured on the BBC's This is Your Life.*

their faith in flesh-and-blood human beings, especially ones who had avuncular smiles, wore glamorous uniforms and had military decorations beneath their wings? BOAC were following a well-trodden path, throughout the Second World War, British propaganda had placed huge emphasis on the courage and skill of pilots. They had good cause, during the conflict RAF Bomber Command had suffered the highest level of casualties of any armed unit: 55,573 killed out of a total of 125,000 aircrew (a 44.4 per cent death rate), with a further 8,403 wounded in action and 9,838 taken prisoner.

BOAC shone a spotlight on their pilots, and some of them became household names. When Captain Oscar Phillip Jones (better known as O P) retired in 1959, his career was featured on *This is Your Life,* a very popular BBC television programme of the day. O P had been a veteran of Imperial Airways and a pioneer of the Empire routes from Cairo. With his military bearing, carefully trimmed beard, cheroot in the corner of his mouth and cap at a jaunty angle, he personified the cool, steady and imperturbable airline pilot in publicity shots from the 1920s onwards. He regularly captained royal flights and when BOAC was promoting their new first class Monarch service,

*Monarch Class in the downstairs lounge of a BOAC Stratocruiser.*

*Former BOAC Chairman, Sir Miles Thomas, would later say of the Handley Page Hermes, 'It had every possible fault except that it didn't blow up!'*

O P was often photographed at the bottom of the steps of a Stratocruiser, welcoming passengers aboard.

Another famous BOAC pilot was Captain J C (Jack) Kelly-Rogers, who flew Churchill across the Atlantic in 1943 and later became Churchill's personal pilot for all transatlantic flights. He had previously commanded Imperial Airways' entire fleet of flying boats and even tested the first inflight refuelling of a flying boat. As late as the 1960s BOAC would continue to feature pilots in their advertising. A television commercial for the VC10 featured a chief steward with the bearing and manner of a senior NCO. He was played by an actor, but the charming captain in the cockpit was played by a real BOAC pilot, Norman Bristow. To the strains of *Kiss me goodnight Sergeant Major*, BOAC projected the aura of an airline that was run with soldierly precision.

The military image was closer to the truth than many realised. Following the Second World War, BOAC had recruited heavily from former armed services personnel who were then being demobbed in large numbers. As might be expected, the majority of pilots, including Norman Bristow, came from the RAF, but many of the ground staff were ex-Army. BOAC, with its uniforms, rank insignia, clear hierarchy and strictly enforced rules was a little like the military and had a similar structure. Senior ground staff were called duty officers, captains were called *sir* by their crews and practically everyone used military jargon to describe BOAC's operations. Most of the flight deck crews had medal ribbons on their jackets and it was not unusual to see DSOs, DFCs, and other important decorations beneath BOAC pilots' wings. Occasionally, you might see a second set of wings which denoted a former Pathfinder pilot. Don Bennett, who had commanded the RAF's elite Pathfinder force, had been the first Managing Director of BSAA and had hired many former Pathfinders as pilots. Most of them transferred to BOAC in 1949.

*Captain J C Kelly-Rogers.*

By the 1960s the supply of experienced, former military pilots was slowing, and BOAC and BEA responded by opening their own pilot-training school at Hamble, near Southampton. The new breed of pilots would be civilians who had chosen airline flying as their career from the outset, rather than RAF men who had drifted into it as their military service came to an end. The Hamble graduates were not always welcomed by older crews, and a joke circulated that they had really been hired as sex therapists, having often be told: 'If I want your f*****g advice I'll ask for it!' Ex-747 Captain Dave Grace had an early experience before he had even reached BOAC: 'In 1968, as a Hamble cadet, I was asked to address the annual technical meeting of the Guild of Air Pilots and Air Navigators

in London. The subject was, "Will today's college-trained pilots be ready for command when their turn comes?" It was a very daunting experience for a young cadet, made no easier by having Captain O P Jones glowering at me from the front row!'

In 1970, BOAC co-operated in the making of a television documentary: *Airline Pilot*, which followed a young Hamble cadet, Stephen Radcliffe, through his pilot training up to his first flight as a second officer on a VC10. At the time he was the youngest pilot in BOAC. The documentary was popular but the 1970s marked the end of an advertising policy based largely on human crews. In the following decades, as technology improved, so the pilots became less visible.

A marketing innovation which paid huge dividends for BOAC was the Junior Jet Club, a working example of the Jesuits' maxim: 'Give me the child and I'll give you the man.' From 1957, children on BOAC flights (myself included) were given a set of gold and enamel pilot's wings, and a Junior Jet Club logbook. The books were beautifully finished in blue with gold leaf, and all your flights with BOAC were recorded in them and signed by the captain. Looking similar to real pilots' logbooks, they showed date, aircraft type and

*A marketing innovation which paid huge dividends for BOAC was the Junior Jet Club.*

registration, departure city, destination city, flying hours and miles. On the flyleaf was a photograph of BOAC's most famous pilot with the message:

*Dear Club member,*

*I should like to welcome you as a joining member of the Junior Jet Club and hope you will be pleased with your own personal logbook. My fellow captains will look forward to meeting you on many future flights and will be delighted to help you build up your BOAC mileage.*

*We should like you to complete the enclosed Enrolment Card and hand it to your steward or stewardess to enable us to keep in touch with you from time to time.*

*Yours sincerely,*

*Captain O P Jones*

It was marketing magic, try telling a twelve-year old boy that he's missed out on several thousand miles in his logbook because you're not flying BOAC! Parents soon discovered they no longer had any choice of airline; they had either to book BOAC or travel with bitterly disappointed kids.

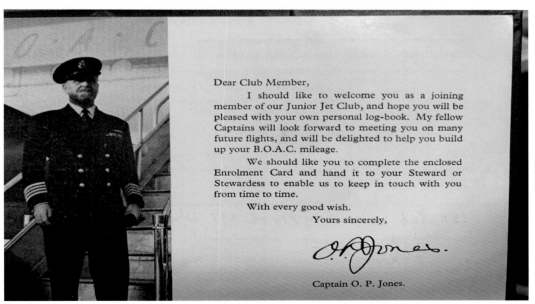

*On the flyleaf was a photograph of BOAC's most famous pilot.*

*Some children accumulated colossal mileages.*

Some children accumulated colossal mileages. BOAC was an exclusively long-haul airline and an important part of its business was flying the children of expatriates to and from the mother country. One of BOAC's specialities was the service it provided for 'Unaccompanied Minors', i.e. children travelling without their parents. This was an era in which British diplomats, armed forces personnel, and overseas managers in British companies often had their children's school fees and travel costs paid as part of their salary. In every school holiday period, thousands of children under the age of eighteen would be chaperoned on BOAC flights by 'Aunties' (female ground staff who volunteered to supervise minors). Such was BOAC's reputation that parents across the world were happy to have their children delivered to London Airport into the sole hands of BOAC, knowing that they would arrive safely at their destination, and be taken back again when the holidays were over.

Retired BOAC captain Ray Howell recalls: 'In an era when carrying children was fairly unusual, I was delighted when I used to be presented with their logbooks. It became a fashion for cabin crew to sign on behalf of the captain, but I liked signing them myself. I saw no reason to use obscure signatures – what point is a completely illegible scribble? I used to sign so my name could be read.'

Former flight engineer Colin Patience noted: 'BOAC were very much involved in the lives of "Children of the Empire", even after the independence of those countries. Thousands of kids flew from Australia and New Zealand, the Far East, Malaysian Peninsula, Indian Subcontinent, Middle East, Africa (North, South, East and West) the Caribbean, as well as South and North America. I knew and have subsequently met people who ended up in schools in UK and Ireland who came from all these places.'

To this day, there are thousands of travellers who have retained and treasure their logbooks, and some who still ask captains of other airlines to sign them. Alasdair Cook is one of them: 'My first flight ever was in July 1961 on a BOAC Bristol Britannia from London to Montreal. It took about twelve hours and we left from the prefab buildings just off the A4 road. I think some of them are still standing. I still have my Junior Jet Club logbook which I had with me recently when I was on my way to join a cruise ship in Barbados and the captain of our TUI Dreamliner signed it for me!'

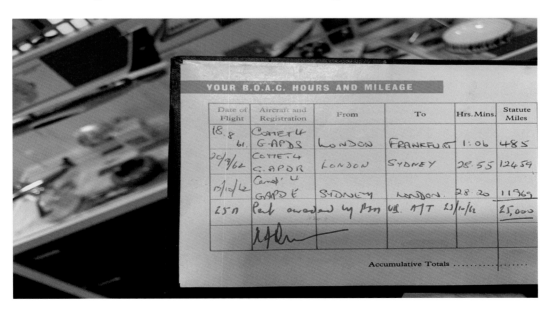

# CHAPTER NINE

## Royalty

Imperial Airways had long understood the value of publicity that came from carrying royalty. In April 1931, the Prince of Wales and his brother the Duke of York (later King Edward VIII and King George VI respectively), flew home from Le Bourget Airport in Paris aboard an Armstrong Whitworth Argosy named *City of Glasgow* (G-EBLF). The aircraft was able to deliver them directly to their door by obligingly landing in Windsor Great Park. The robust, fixed landing gear and enormous wheels of the hefty biplane meant there was no problem landing on grass.

In 1952, Princess Elizabeth and her husband, the Duke of Edinburgh, had travelled to Kenya at the beginning of what had been planned as a five-month colonial tour. The wind of change had begun to blow through the Empire and British rule was under pressure in Kenya. Mau Mau fighters were gaining ground in their struggle

*In 1931 Armstrong Whitworth Argosy G-EBLF delivered the Prince of Wales to his door by landing in Windsor Great Park.*

*King George VI sees off his daughter, Princess Elizabeth, at London Airport in 1952.*

for independence (they eventually won in 1963), and the King, who was too ill to travel himself, felt it was necessary for royalty to be seen in Africa.

Years later, Lord Chandos, who had been Colonial Secretary at the time, would tell *The Daily Telegraph*: 'I well remember the last time I saw the King. When Princess Elizabeth and Prince Philip left Heathrow for Kenya, the King and Queen came to see them take off... I was shocked by the King's appearance. I was familiar with his look and mien, but he seemed much altered and strained. I had the feeling of doom, which grew as the minutes before the time of departure ebbed away. The King went on to the roof of the building to wave goodbye. The high wind blew his hair into disorder. I felt

*The royal party was taken aboard a hurriedly chartered East African Airways DC3 (Sagana), which flew them from Nanyuki in Kenya's Rift Valley to Entebbe, in Uganda.*

with foreboding that this would be the last time he was to see his daughter, and that he thought so himself.'

Five days into the royal trip, King George VI died in his sleep at Sandringham. The renowned big-game hunter Jim Corbett, who had been staying with the royal party at the Treetops game lodge in the Aberdare Forest, wrote these lines in the visitors' guest book: 'For the first time in the history of the world, a young girl climbed into a tree one day a Princess and after having what she described as her most thrilling experience, she climbed down from the tree next day a Queen.'

Rapid arrangements were made to bring the new Queen home to England. Storms were forecast and for a while it was not clear whether the BOAC Argonaut (G-ALHK Atalanta) would be delayed. The captain, Ronald Ballantine, was an Imperial Airways veteran who had been nicknamed 'The Colonel' because General Chiang Kai

Shek once offered him a colonelcy in his nationalist air force – a post which Ballantine had politely declined.

The royal party was taken aboard a hurriedly chartered East African Airways DC3 (*Sagana*), which flew them from Nanyuki in Kenya's Rift Valley to Entebbe, in Uganda. There they would have to wait in the airport restaurant for the tropical storm to abate. By the time the BOAC Argonaut, with Ballantine at the controls, had arrived from Mombasa, the wind speed had reached 50 knots and Ballantine decided that the take-off from Entebbe would have to be delayed. Purple and yellow lightning flashed continuously over the aerodrome, dimming the runway lights and illuminating the thrashing flag which flew at half-mast from the control tower. Sheets of rain lashed the silent planes and bounced off the black tarmac of the runway.

A room was prepared for the royal couple in the airport manager's private office, but they preferred to stay in the restaurant, where the Queen drank a soft drink. For an hour the storm raged at its height, but then slowly subsided. The Argonaut finally took off just before noon and arrived at London Airport on the following evening, to be met by the prime minister, Sir Winston Churchill, along with Clement Atlee, and Anthony Eden, on the rain-glazed tarmac of a cold, grey February day.

The unexpected death of the King had seized the imagination of the public like no other event since the war. In an age where television was beginning to challenge print and radio (there were then 1.5 million sets), the images of the new Queen descending the steps of a BOAC airliner became engraved in the national psyche. Like Chamberlain's return to Heston in 1938, the newsreels would be repeated endlessly down the decades. BOAC was now more than an airline; it was a part of our national heritage.

*The Queen, then Princess Elizabeth, names an Avro Tudor aircraft Elizabeth of England by pouring English cider onto its nose in 1947. However, despite the fanfare, it was not to enter service with BOAC who later cancelled their order.*

# CHAPTER TEN

## The Jet Age

In 1952, BOAC would pioneer the single greatest technical advance in airline history, to be followed by tragedy and failure within just two years.

Frank Whittle's jet engine and Alan Turing's computer had been Britain's outstanding technical breakthroughs during the Second World War. Because the Air Ministry had been slow to recognise the potential benefits, viable jet fighters were not delivered to the RAF in time to alter the course of the war. Nonetheless, by the early 1940s the de Havilland company, who had always been among the most imaginative and forward-thinking of aircraft designers, had realised that jets could transform civil aviation.

As with other developments, the Brabazon Committee had not been quick to see the commercial opportunities of jet propulsion. However, Sir Geoffrey de Havilland, who was both a Committee member and head of the de Havilland company, enthusiastically

*The prototype de Havilland Comet - the world's first jet airliner.*

pushed for a pure turbojet-powered design. The Committee took some persuading, but finally accepted his proposal and awarded a development and production contract to the Handley Page company. It soon became clear that Handley Page were struggling with the challenge and in February 1945 the project was passed to the de Havilland company, who took on the formidable task of designing and building both the airframe and engines of a new jet airliner. Initially it was to be a small mail-carrying aircraft with seats for just six passengers, but that soon evolved into a long-range airliner that would carry 24. Showing extraordinary courage and foresight given the challenges involved, BOAC placed an order for 25 (subsequently reduced to ten), and the Jet Age began.

By 1949 the first prototype DH106 Comet airliner was complete. Like earlier de Havilland aircraft, including the Mosquito and the air race winning DH88 (which had also been named Comet), she was ravishingly beautiful with clean, sleek lines, a swept wing and elegant square windows. On 27 July she made her first flight from Hatfield under the command of de Havilland's Chief Test Pilot, John 'Cat's Eyes' Cunningham, a war hero who had earned his nickname

*A cutaway view of the de Havilland DH106 Comet which first flew in 1949.*

in night fighters. Two months later she made her first public appearance at the Farnborough Air Show, at which point public and press interest became intense. Not only de Havilland, but BOAC and the government of the day had a great deal riding on this revolutionary new plane. The future of both the British aircraft industry and the British airline industry, along with a large dose of much-needed national prestige (post-war austerity was still the order of the day) rested upon her swept, silver wings.

By 1951 BOAC had received their first production aircraft, G-AL-YP, and route-proving flights began. On 2 May 1952, *Yoke Peter* took off from London Airport on the world's first scheduled jet service with fare-paying passengers. Newsreel cameras were on board to record the flight to Johannesburg with stops at Rome, Beirut, Khartoum, Entebbe and Livingstone before landing at Jan Smuts Airport nearly 24 hours later.

The gamble taken by de Havilland, BOAC and the government appeared to have paid off in spades. Flying times were literally cut

*On 2 May 1952, Yoke Peter took off from London Airport on the world's first scheduled jet service with fare-paying passengers.*

in half and in their first year alone Comets carried 30,000 passengers. BOAC quickly discovered that their new aircraft could be profitable with a load factor as low as 43 per cent, and they were fuel-efficient above 30,000 ft. By the summer of 1953, eight BOAC Comets were leaving London each week: three to Johannesburg, two to Tokyo, two to Singapore and one to Colombo. BOAC was not only making headlines; they were making money too and de Havilland, who had a five-year lead on the rest of the world, were booking orders from foreign airlines.

*Cabin service aboard a BOAC Comet.*

But, like a Greek tragedy, nemesis would follow and BOAC's days in Arcadia would quickly come to an end. Just two months after its introduction a BOAC Comet crashed during take-off from Rome's Ciampino airport, having failed to get airborne. Although there were no fatalities or serious injuries, the airframe was a write off. A few months later a Canadian Pacific Comet failed to get airborne from Karachi, and this time there were ten fatalities. Initially, pilot error was thought to be the cause of both accidents, although de Havilland later discovered a flaw in the Comet's design which allowed the wing to stall while still on the ground if the pilot raised the nose too high. They subsequently modified the wing.

In May 1953 a BOAC Comet broke up in a thunderstorm six minutes after taking off from Calcutta and all 43 souls on board perished. Over-controlling of the aircraft (the Comet was the first airliner to have fully powered controls), was blamed and more modifications were made. No one considered metal fatigue as a possible cause. Up until this point pilot error, weather and a certain amount of bad luck were still thought to be the culprits. In an era when aircraft accidents were more frequent than today, few suspected that there was a fundamental problem with the aircraft's design.

In January 1954, disaster struck again. G-ALYP, BOAC's first Comet, broke up in mid-air 20 minutes after taking off from Rome with 35 fatalities. Media attention focussed on possible sabotage, but BOAC decided to ground its entire Comet fleet as a precaution while investigations continued. A committee looked at six different possible causes, including metal fatigue of the wings, but finally concluded that fire was to blame. The wreckage had fallen into the sea near Elba, and the Royal Navy were slowly recovering it, but before it could be properly examined the committee announced that there was no apparent fault with the Comet, and the government decided against a full public enquiry. There can be little doubt that national prestige and the financial implications for both BOAC and de Havilland played a role in this decision.

Comet flights resumed on 23 March 1954, but just two weeks later Comet G-ALYY, which had been leased to South African Airways, broke up in mid-air, once again on departure from Rome. This time 21 lives were lost, and all Comets were immediately grounded. Prime Minister Churchill ordered a full enquiry and the Royal Aircraft Establishment at Farnborough eventually proved that metal fatigue, brought about by repeated pressurisation stresses and mis-drilled holes, had caused the Comets to break up. There were a number of faults with the Comet's design, but the principal one was that the fuselage simply wasn't strong enough – largely the result of insufficient knowledge about the distribution of stress in a pressurised airframe. The failures had occurred around rivets and bolt holes, but a myth soon sprang up that the Comets had crashed because they had square windows. Although subsequent Comets had round windows, the earlier square examples had little to do with the accidents and the same shape was used successfully on other pressurised airliners, including the Boeing 377 and Douglas DC7. The Comet 1 never returned to service and BOAC had to bring older piston-engined aircraft back onto their routes. Worse still, they were forced to buy up second-hand Constellations and Stratocruisers at inflated prices, and it would be four years before they could reintroduce jets.

The legacy of the Comet disasters reverberates to this day. The British aircraft industry would never build a commercially successful long-haul airliner, and in 2002 would cease to build complete airliners altogether. Concorde would be a resounding technical success but a commercial failure. In 1957, de Havilland tried for a final time with the short-to-medium-range DH121 Trident. A technically superb design, it might have been a world-beater had BEA not insisted on reducing both the size of the aircraft and the powerplant.

BOAC took a massive financial and reputational hit. Their reputation would recover in time, but the financial losses would accumulate until they were written off by the government during the

*In 1958 the completely redesigned and much improved Comet IV entered service and BOAC could bring the world back into the Jet Age.*

*A BOAC Comet IV lands at Hong Kong.*

early 1960s. In the United States, Boeing and Douglas would learn from de Havilland's mistakes and both would produce successful long-haul jets, especially the Boeing 707, for which BOAC itself would be an important customer. In 1958 the completely redesigned and much improved Comet IV entered service and BOAC could bring the world back into the Jet Age. Pan American were about to receive their first Boeing 707s and BOAC knew they would recover some of their lost prestige if they could be the first airline to fly passengers across the North Atlantic, which had long been the most glamorous and profitable of their routes. To say they moved quickly would be an understatement: the first two Comet IVs were handed over to BOAC at London Airport on 30 September and just five days later they would simultaneously fly passengers between London and New York, one operating westbound and the other eastbound.

The west-to-east flight, (G-APDB) commanded by Captain Thomas Butler (Tom) Stoney, DFC, departed Idlewild Airport in New York at 7:01 a.m., local time, with Basil Smallpiece, and Aubrey Burke,

*BOAC inaugurates the first transatlantic jet service in 1958.*

managing directors of BOAC and de Havilland, respectively, on board. Benefiting from favourable winds, the eastbound flight took just 6 hours, 12 minutes, averaging 565 miles per hour.

The east-to-west airliner, G-APDC, departed London Airport at 8:45 a.m., London time, under the command of Captain R E Milli-chap, with Sir Gerard d'Erlanger, chairman of BOAC, and 31 other passengers aboard. The westbound flight took 10 hours, 20 minutes, including a 1 hour, 10-minute fuel stop at Gander Airport, Newfoundland. Both airliners were configured to carry 48 passengers.

BOAC reaped a huge public relations and publicity dividend, but it would be short lived. Juan Trippe, BOAC's eternal competitor, was hard on their heels. Three weeks later a Pan American Boeing 707 would cross the Atlantic from New York to Paris and open a new and different chapter in civil aviation. The Boeing 707, which was larger than the Comet and could carry (according to its configuration) a lot more passengers, significantly reduced air fares. The 707 not only made air travel faster and more comfortable, it also made it more affordable and brought long-haul flying within the budget of the average man for the first time. It was the 707, and not the Comet, which would change the world in the following decade. De Havilland's beautiful pioneering creation – a national project in which the British had invested so much money, effort, time and pain, was already obsolete. By 1962 BOAC began retiring their Comets and replacing them with the Boeing 707s that they had begun ordering as early as 1956. The press and the trade unions accused BOAC of betrayal, but the 707 made a lot more money for them than the Comet.

Not everybody saw the increase in speed as a benefit. In the James Bond novel *For Your Eyes Only* (1960) Ian Fleming lamented: 'Two days later, Bond took the Friday Comet to Montreal. He did not care for it. It flew too high and too fast and there were too many passengers. He regretted the days of the old Stratocruiser – that fine lumbering old plane that took ten hours to cross the Atlantic. Then one had been able to have dinner in peace, sleep for seven

*Bob Leedham, flight engineer on the first scheduled transatlantic crossing by jet aircraft, back on the flight deck of the BOAC Comet 4 at Duxford.*

105

*Test pilots Brian Trubshaw (right) and Gabe 'Jock' Bryce (centre) following the first flight of the VC10.*

hours in a comfortable bunk, and get up in time to wander down to the lower deck and have that ridiculous BOAC "country house" breakfast while the dawn came up and flooded the cabin with the first bright gold of the Western Hemisphere.' There was logic in this, wealthy passengers like Fleming generally had plenty of time, they were more concerned with comfort and luxury than with speed.

Despite the success of the 707, the British aircraft industry was not yet ready to throw in the towel. During the 1950s the Vickers company had designed a military transport aircraft called the V1000, which was a derivative of their Valiant bomber. They also planned a civil airliner version named the VC7. At a late stage in the aircraft's development the RAF cancelled their order and BOAC turned down the civil version. George Edwards, Vickers' outspoken managing director commented: 'We have handed to the Americans, without a struggle, the entire world market for big jet airliners.'

BOAC were happy with the 707 for their North Atlantic routes, but in 1957 they called for a new type of aircraft to replace their Britannias and Comets on the African and Asian routes which the 707 was less well-suited to. Vickers answer was the VC10, a four-engined, long-range aircraft that would able to carry a useful payload into the 'hot and high' airports that had long been a problem. Instead of the under-wing engine configuration of the 707 and DC8, Vickers opted to cluster four Rolls-Royce Conway engines (the same powerplant that BOAC had specified for their 707s) at the rear. The advantage of this design was a 'clean wing' which, unencumbered by engines and pylons, reduced the take-off and landing speed by around ten knots, allowing the VC10 to operate from airports with high elevations and short, narrow runways. The disadvantage of the design was that the necessary strengthening of the wing and tail increased the weight and fuel consumption, thereby making the VC10 less economical than the Boeing. BOAC placed an order for 35, with options for 20 more, although the numbers would change several times before the aircraft were delivered.

In 1963 the government appointed Giles Guthrie as Chairman and Chief Executive of BOAC. Although the airline was then making a profit, it had made heavy losses in the late 1950s and early 1960s and when Guthrie arrived it was carrying a total debt of £80 million (every penny of which was owed to the government). BOAC also had

a capital borrowing facility in place for a further £180 million which had been set aside to pay for the VC10s. Guthrie had been hired because he had a background in both aviation and banking, he was a distinguished pilot who had previously been a director of BEA and had also run his family's merchant banking business. He was the first head of BOAC to be told that he must run the corporation along purely commercial lines, i.e. for profit, and not as a loss-making public service. The government had made it clear that they wanted to see BOAC returned to long-term profitability.

The plan that Guthrie presented to the Minister of Aviation, Julian Amery, was bold and simple: the only way that BOAC could be put back on its feet financially was for the government to write off all the debt and refinance BOAC as if it were a new start-up business. At the same time, the VC10 orders should be cancelled. Amery agreed to write off the debt but not to cancel the VC10, fearing that this would destroy the British aircraft industry. As a compromise, Guthrie agreed to reduce BOAC's order for VC10s. BOAC would now operate a fleet of both VC10s and 707s in roughly equal numbers.

Guthrie's plan worked and BOAC returned to growth and profitability. He resisted government pressure to purchase a proposed

In the end, thanks to brilliant marketing by BOAC, the VC10 was a commercial success.

double-deck variant of the Vickers Super VC10, and by August 1966, he had placed orders for the world's first wide-body aircraft, the Boeing 747. He won this argument largely because, as had so often happened in the past, it was known that Pan American would operate the 747 on the North Atlantic, meaning that BOAC would have little option but to follow suit. All the same, there were elements in BOAC, and elsewhere, who would never forgive him for trying to stop the VC10. Caroline Ely, who was a BOAC stewardess at the time recalled: 'I remember that there were notes appearing on the flight deck which said, "Put Guthrie in Khartoum", that's what people thought of him.'

In the end, thanks to brilliant marketing by BOAC, the VC10 was a success, becoming so popular with passengers that on some routes the load factors were anything up to 50 per cent higher than for the rival 707. Like so many other British airliners, including the earlier Britannia and Comet IV, the VC10's problems had really been caused by delay. By the time it went into service the Boeing 707 had been operating for six years, the DC8 for four, and most of the world's long-haul airlines had already opted for one or the other.

On some routes VC10 load factors were anything up to 50 per cent higher than for the rival 707.

Worse still, by the early 1960s, the 'hot and high' runways for which it was specifically developed had been lengthened to accommodate the 707 and DC-8, depriving the VC10 of its unique selling point.

The blame for the failure of the British aircraft industry to develop a commercially successful long-range airliner can be traced all the way back to the shortcomings of the Brabazon Committee in the 1940s. At the same time, BOAC's huge losses prior to Guthrie can, to some degree, be laid at the door of BOAC's first chairman, the redoubtable Lord Reith. His insistence that Imperial Airways, and later BOAC, should be run as a public service and not for profit had hamstrung the airline from the beginning. Unlike the BBC, an institution which Reith had created in his own image, there was no public appetite for a subsidy. While voters might accept an annual licence fee in exchange for advert-free radio and television, they weren't going to tolerate taxpayers' money being poured into the bottomless coffers of loss-making airlines. These arguments would rumble on until British Airways was privatised and floated on the London Stock Exchange in February 1987.

*A BOAC 707 at Sydney Airport.*

# CHAPTER ELEVEN

## The Sixties

During the twentieth century there had been two decades, both of which came in the aftermath of world wars, in which sweeping economic and social change had led to new perceptions of almost everything. The first was the Jazz Age (or Roaring Twenties), and the second was the 1960s. They helped to define both Imperial Airways and BOAC.

Imperial Airways had been born in the *entre-deux-guerres* period, a time of hedonism, wealth, freedom, and youthful exuberance which had been iconised in the seminal novel, *The Great Gatsby* (1925), by F Scott Fitzgerald. Changes in cinema, writing, painting and architecture had helped shape the public's perception of air travel. Ex-First World War aviators barnstormed their way across

*Julie Christie in the 1965 film Darling.*

America. Hollywood presented pilots as daredevils and heroes in films including *The Flying Ace*, and *Flight*. Amelia Earhart broke records and Charles Lindbergh, America's Lone Eagle, crossed the Atlantic to become an overnight sensation. Imperial Airways exploited the zeitgeist to produce brilliant posters and advertisements that are collected to this day. Flying was seen as daring, fashionable, modern and exclusive – Imperial benefited from all those associations.

During the Second World War, air transport would again become a strategic and military necessity while virtually all civilian flying was suspended. But as prosperity returned to Britain from the late 1950s, another huge change took place. The Jet Age not only shrank the world, it democratised air travel by making it available to the common man and, as it did so, perceptions of airlines altered. BOAC was more than a bystander while this happened; they played a key role in changing the way we all saw international travel.

As we approach the third decade of the twenty-first century, we have come to mistrust technology and think of it, at best, as a

*In the summer of 1951, the Festival of Britain was a celebration of all that was new and technical.*

double-edged sword. Climate change, pollution, and the loss of natural environments have made us cynical about technical 'progress' and fearful for the future. But in the summer of 1951, millions of people gathered in London for a celebration of all that was new and technical. The Festival of Britain had as its centrepiece the Skylon: a futuristic sculpture which looked rather like an aerodynamically shaped rocket and appeared to float above the ground. Historian Kenneth O Morgan enthused: 'People flocked to the South

Bank site, to wander around the Dome of Discovery, gaze at the Skylon, and generally enjoy a festival of national celebration. Up and down the land, lesser festivals enlisted much civic and voluntary enthusiasm. A people curbed by years of total war and half-crushed by austerity and gloom, showed that it had not lost the capacity for enjoying itself. Above all, the Festival made a spectacular setting as a showpiece for the inventiveness and genius of British scientists and technologists.' Among them were the scientists and engineers who had made the Comet possible and would later build Concorde. This would be an era in which new technology and new ideas would be celebrated.

In January 1958 Frank Sinatra sang, 'Come fly with me, let's fly, let's fly away...' On the album cover he smiled and cocked his thumb towards a TWA Lockheed Constellation, but he was already behind the curve. By October of that year BOAC and Pan Am would put jets into service and banish the old piston-engined liners to history.

In November 1961, the long-anticipated Oceanic Building (today's Terminal 3) opened in the Central Area of London Airport. Designed by Sir Frederick Gibberd in the Modern Movement style (he had been the architect of Harlow, the first post-war New Town), it was a quantum leap from the wartime prefabs BOAC's passengers had endured before. Its principal features were a very high ceiling, ground to roof glass and what seemed like acres of empty floor space. Huge modern art murals were mounted in the windows while wood-panelled walls and wide staircases added to the sense of grandeur. Glass cabinets displayed the luxury goods that could be bought

*In November 1961, the long-anticipated Oceanic Building (today's Terminal 3) opened.*

*This was an era in which passengers wore smart clothes to travel.*

in duty-free and the widely spaced check-in desks sported the logos not only of BOAC, and Pan Am, but also of the many new national airlines that now competed on the old Empire routes. Inside, passengers were surrounded by cathedral-like calm broken only by an occasional tannoy announcement. Queues were either short or non-existent and, except during weather delays, the building was rarely crowded. This was an era in which passengers wore their best clothes to travel and the airport was part of the adventure – photographs were taken, friends and relatives came to see you off. The roofs of the nearby Queen's Building were a maze of balconies on which hundreds of visitors, for a small fee, would spend an entire day watching the aircraft arrive and depart. The new Oceanic Building was a triumph of style over function and a celebration of the Jet Age; it had been built to impress, rather than simply channel the passenger from their taxi to the aircraft door.

As the 1960s unfolded so England, and Liverpool and London in particular, became the epicentre of social change. Liverpool had long been Britain's North Atlantic port and in the 1950s merchant seamen had brought home records by new American artists which could not then be bought in British shops. Local groups began to produce a different kind of music strongly influenced by American greats including Chuck Berry and Buddy Holly. Since the Second World War British popular culture had largely followed that of the USA, but that was about to change. By 1964 the Beatles and Rolling Stones had exploded onto the music scene and the 'British Invasion' of the United States had begun. London was beginning to swing, and BOAC would reap huge rewards. Between 1960 and 1970 the number of passengers carried by BOAC almost quadrupled.

As London became cooler, so it became a more popular destination, especially for Americans. In February 1964 the Beatles made

*Twiggy and boyfriend Justin de Villeneuve departing from London in the 1960s.*

their first tour of the United States. BOAC's marketing department rarely missed out on these kinds of opportunities but, not for the first time, they were outpaced by a more agile Juan Trippe. It was Pan Am that flew the Fab Four to New York aboard their 707 Jet Clipper *Defiance*. They were met at Idlewild Airport by 4,000 fans plus 200 journalists and their two appearances on the Ed Sullivan Show were watched by 70 million, a record at the time. Their tour of America established

*George Harrison's hair is combed by a BOAC stewardess.*

them as the most popular music group of all time and changed cultural history on both sides of the Atlantic. When they returned to London at the end of the month, British Pathé News called it, 'The greatest day that London Airport had ever known' – 12,000 screaming fans filled every terrace and balcony of The Queen's Building to welcome them home. This time, the Pan Am 707 had been christened (with a hastily added sticker) Jet Clipper *Beatles*. The group would later use BOAC for other tours including one to Australia, but nothing would eclipse the impact of the Beatles' first American tour.

BOAC must have been stung by their failure to carry the Beatles, but they did not miss an opportunity to fly Frank Sinatra, the biggest international star of the pre-Beatles era. George Cleaver's father, Terry Cleaver, was a public relations officer for BOAC in the Far East, stationed in Hong Kong. George recalls how it happened: 'One day dad received a call at the office from his friend, a Pan Am public relations officer, inviting him and my mother to dinner with Frank Sinatra and Leo "Lippy" Durocher, the celebrated baseball player. They were passing through Hong Kong with their wives on their way back to New York, via London. During the dinner dad asked Frank if he'd ever flown BOAC to which he replied, "No, Terry, you guys don't carry Jack Daniel's Green Label on your flights." Dad responded, "Well, Frank, if we can find some of your favourite whiskey, would you consider giving us a try back to London?" When Sinatra agreed, Dad turned to his Pan Am friend and asked if he'd mind comping their tickets over to BOAC, to which he received a wink of approval. The next day dad arranged a plenitude of Jack Daniel's Green Label for the flight and was graciously received by the flight crew for putting such distinguished guests on their aircraft.'

By the late 1960s a visit to London had become almost obligatory for any young American who wanted to be considered part of the zeitgeist, and BOAC became the airline of choice. New excursion and advanced purchase fares were making air travel increasingly more affordable and BOAC was gaining a level of brand recognition it had never known before. In an age where cheap silkscreen printing meant that graphics had become pervasive, from Andy Warhol's Campbell's Soup tins to the stylised Union Jacks that appeared everywhere, BOAC's gold Speedbird had become the world's most recognised airline logo.

*Diana Dors and her first husband arriving by BOAC.*

For crews, as well as passengers, the 1960s would be a golden age. BOAC's routes meant long stopovers in exotic destinations, often staying in fine hotels. In Asia and Africa, they were provided with full board, but in the USA, they were given cash allowances in US dollars. The crews who flew regularly to North America became known as the Atlantic Barons. Some hotels could be basic, in Karachi for example, BOAC had its own accommodation known as The Speedbird Guest House which was not always popular. As retired steward Steve Frampton recalled: 'The Speedbird Guest House in Karachi was a fascinating place. Rooms were like prison cells complete with Flit cockroach spray pumps!' Other hotels, however, were famous for their luxury including the Peninsula in Hong Kong and Raffles in Singapore. In some cities, flight deck and cabin crew were separated. In Nairobi for example, flight deck crew stayed at the plush Stanley Hotel in the city centre, whereas cabin crew were put up at the more basic Grosvenor, near to the airport.

*On Japanese flights BOAC used locally employed cabin crew who wore kimonos.*

As prosperity grew so BOAC expanded its route network to reach more exotic destinations. Jamaica, the Bahamas and Bermuda had long been served, but as more Caribbean islands acquired modern airports, including Barbados, Antigua and St Lucia, so BOAC would begin to fly to them, sometimes contributing to their costs and often creating a luxury tourist industry almost overnight. Until 1972 the Seychelles could only be reached by sea. When BOAC began to fly to these exquisite and unspoiled Indian Ocean islands, it quickly turned the destination into one of the world's most exclusive resorts.

*Comet 4 G-APDE at Vnukovo Airport, Moscow 20th Feb 1959, delivering Harold Macmillan to Moscow. (Photo courtesy of John Furlong).*

The Cold War had made it difficult for BOAC to operate into, or even over the USSR (there was an exception in 1959 when a BOAC Comet IV took Harold Macmillan to Russia for talks with Nikita Khrushchev). But in 1970 BOAC finally got permission to operate their London to Tokyo service via Moscow. One of the first crews to stop overnight there had joined the captain for a drink in his hotel room. The talk turned to the KGB who were believed to routinely bug foreigners' hotels. A search of the room began, and they eventually found a suspicious looking grey metal disk hidden beneath the carpet. It was held in place by several stout screws and a bolt which the flight engineer decided to remove using some tools that he always kept in his flight bag. When the screws were finally out and the bolt loosened, they heard a deafening crash from the floor below. There was now a hole in the floor where the disk had been and as they peered through it they saw a group of stunned (but thankfully uninjured) guests standing around the remains of an enormous chandelier which had shattered into a thousand pieces across the lounge floor.

While BOAC grew, the British people were being tempted to take their holidays on the Continent by a new breed of charter airline who were making cheaper package tours possible. Dan-Air were using most of BOAC's old Comet IVs from Gatwick, while Britannia Airways (formerly Euravia) and Monarch Airlines were operating ex-BOAC Britannias, and Court Line (formerly Autair), was flying a brightly coloured fleet of BAC 1-11s to popular destinations from Luton. At the same time, Freddie Laker, who had been the managing director of the independent British United Airways, was forming his own company, Laker Airways, with a fleet of two ex-BOAC Bristol Britannias. Laker already had his eye on BOAC's lucrative North Atlantic routes. By the late 1960s several other charter airlines, including Donaldson International Airways, Caledonian Airways, and Lloyd International Airways had bought second-hand Boeing 707s which they were operating across the North Atlantic by exploiting

the relatively loose charter rules which were in force at the time.

In 1960 the shipping line, Cunard, had acquired a stake in the independent airline, British Eagle, which enabled Eagle to buy two new Boeing 707s, also destined for the North Atlantic. BOAC immediately saw the new company, Cunard Eagle, as a direct threat to their dominance on transatlantic routes and began secret talks with Cunard. In 1962, the two companies formed BOAC-Cunard Ltd to operate scheduled services to North America, the Caribbean and South

America. BOAC provided 70 per cent of the new company's capital and eight Boeing 707s. As part of the deal, Cunard Eagle was dissolved and their two 707s were handed to BOAC. Without Cunard's investment, British Eagle became unprofitable and later collapsed into insolvency. The new marriage did not last long, by 1966 BOAC-Cunard had been dissolved and BOAC continued alone. It is difficult to see what they had gained from the Cunard link, other than the elimination of a competitor.

One of the two 707 aircraft that BOAC had acquired from British Eagle, G-ARWE, was later destroyed in a tragic accident following an engine fire, which claimed five lives including that of BOAC stewardess, Barbara Harrison. She was posthumously awarded the George Cross (the only one ever presented to a woman in peacetime) for her courage in trying to rescue a disabled passenger from the burning aircraft. Two others received awards for heroism: BOAC's Neville Davis-Gordon was awarded the British Empire Medal and an air traffic controller, John Davis, received an MBE. As a result of the accident, BOAC changed the checklists for severe failures and engine fires, combining them both into one checklist.

*Without Cunard's investment, British Eagle became unprofitable and later collapsed into insolvency.*

# CHAPTER TWELVE

## The Jumbo Jet

As early as 1963, the United States Air Force had begun a series of study projects on a very large transport aircraft. The new plane would need to carry close to a hundred tons for distances of up to five thousand miles but have no more than four engines. This was a huge technical challenge for the engine and aircraft makers of the day, but Boeing, Douglas and Lockheed all submitted designs for the airframe. Lockheed won the contract with their C5 Galaxy, but Boeing came up with a fresh idea: with the airline business growing rapidly, could such a large aircraft be used to carry fare-paying passengers? After all, Pan Am's Juan Trippe, who was becoming concerned about airport congestion, had already asked them if they could build a plane twice the size of the 707. Boeing took a spectacular gamble: in the full knowledge that failure would almost

*A BOAC Boeing 747-136.*

certainly bankrupt them, they agreed to take on $2 billion dollars of debt and went ahead with the Boeing 747.

In April 1966, Pan Am ordered 25 747-100 aircraft for US$525 million. During the contract-signing banquet in Seattle on Boeing's 50th Anniversary, Juan Trippe predicted that the 747 would be: '... a great weapon for peace, competing with intercontinental missiles for mankind's destiny'. As the launch customer, and because of its early involvement before placing a formal order, Pan Am was able to influence the design and development of the 747 to an extent unmatched by a single airline before or since. Once again, Trippe had set the bar very high and BOAC felt they had little option but to follow. They ordered eleven aircraft. Had this been a mistake, it might have destroyed BOAC, but, despite teething troubles the 747 would become the most commercially successful aircraft in the airline's history. So much so that half a century later, British Airways is the world's largest operator of the Boeing 747 with 36 examples still in service, one of which has been retro painted in the colours of BOAC!

On 15 January 1970, First Lady of the United States Pat Nixon christened Pan Am's first 747, at Dulles International Airport (later Washington Dulles International Airport) in the presence of Pan Am chairman Najeeb Halaby (Juan Trippe had retired in 1968). Seven

*The arrival of a Boeing 747-136 of BOAC, the first wide bodied aircraft to land at Manchester.*

*First class on BOAC's Boeing 747.*

days later the aircraft flew passengers from New York to London. BOAC received its first three Boeing 747s in May of the same year, but they remained firmly on the ground for months while the corporation argued with its pilots over crewing requirements.

This was not the first time that BOAC had fallen out with the pilots and their union, the British Airline Pilots Association (BALPA). In 1924, Imperial Airways had spent the first three weeks of its existence grounded by a pilots' strike, and relations with its pilots had always been uneasy. BALPA was established in 1937 by Imperial Airways pilot Eric Lane-Burslem, who, when flying an ice-laden four-engine biplane, had a narrow escape after all four engines cut out at 9,000 feet. This, along with another serious incident two years earlier, persuaded Lane-Burslem to form a pilots' association. The first official mass meeting of pilots was held on 27 June 1937.

In 1952, when a BOAC Comet crashed taking off from Rome, the captain had not only been blamed, he had been forced to sign an admission of guilt and demoted to flying the Avro York, then the slowest aircraft in BOAC's fleet. Eight other Comet pilots had been disciplined for accidents during the brief period before de Havilland acknowledged there was a design flaw and modified the Comet's wing. These incidents had led to a feeling among BOAC pilots that they would always be blamed for accidents and that did little to improve relations with management. Additionally, BOAC pilots' pay had fallen some way behind that of American and many European pilots. Retired BOAC pilot Ian Frow reminisced: 'As for the rewards of the job for my first year of employment – spent mostly on the ground doing a mini university course on navigation, BOAC paid me just £50 a month – as a "cadet pilot". Once qualified, the salary rose slowly but steadily with seniority and rank. However, when, after eighteen years I achieved my first command on the Boeing 747, my salary did not exceed £7,000 per annum, by Government command.'

In the late 1960s a first officer on Boeing 707s, Norman Tebbit, became a shop steward for BALPA and began to negotiate for better pay and conditions. Tebbit, who later held several cabinet roles in Margaret Thatcher's government, helped organise a pilots' strike in 1968 which completely grounded BOAC for several days, but

*Former BOAC pilot and BALPA official, Norman Tebbit, who lat-
er held several cabinet roles in Margaret Thatcher's government,
helped organise a pilots' strike in 1968.*

... this should come as no surprise    ➤ **BOAC**

*BOAC pilots took part in route-proving flights and test flew Con-
corde at supersonic speeds.*

ultimately led to pilots getting more money and other privileges, including the right to 'bid' for their favourite routes. A long-standing demand of BALPA had been for a minimum of three pilots on the flight deck, and when BOAC proposed to operate the 747 with only two, BALPA refused to fly it. It took BOAC close to a year to resolve the dispute. During the time that the first three jumbos were grounded, their engines were leased to Pan American.

When the 747 finally went into operation, BOAC faced a challenge as great as the introduction of the first jets in 1952. In well under two decades the passenger capacity of a BOAC flight had increased from 24 to 360, a factor of 15! BOAC's marketing and sales teams were going to have to find massively more passengers in a very short space of time. The pressure was made worse by the fact that a half-empty 747 burned 90 per cent of the fuel of one that was full!

Inevitably, fares plummeted, and passengers soon saw the cost of long-haul air travel falling to the point where holidays in the USA

*British Airways is the world's largest operator of the Boeing 747 with 36 examples still in service, one of which has been retro painted in the colours of BOAC.*

*British Airway's retro painted Boeing 747-400 is met by photogra-phers at London Heathrow.*

or even Australia were now a realistic prospect for those on average salaries. The change was complete, flying was no longer exclusive-ly for the rich and powerful; it was now for everyone. The Golden Age of Flying would have one more dazzling moment before the gilt began to fade. The upper deck of the 747 meant that airlines could bring back the lounges that had once been offered on Stratocruisers and flying boats. BOAC installed a bar and comfortable couches. First class passengers could now climb a spiral staircase to enjoy a cocktail in a relaxing, spacious lounge. But not for long, it did not take airline economists a great deal of time to calculate that filling the upper deck with economy class passengers was infinitely more profitable than giving a handful of first class travellers another perk. The removal of the upstairs lounges coincided with an indus-try-wide process of gradually reducing service while increasing pas-senger capacity, resulting in ever more cramped aeroplanes but, at the same time, ever lower fares. Sadly, for BOAC, a prolonged race to the bottom had begun, and it would not be won by any of the so-called 'Legacy' airlines. The deregulation of the US airline industry in 1978, followed by the European airline industry in 1987, would create a new breed of low-cost carriers including the Dallas-based Southwest Airlines, Dublin-based Ryanair and Luton-based easy-Jet, who would slash fares and eliminate almost all of the complimentary services that passengers had traditionally enjoyed, from free meals to generous luggage allowances.

Nevertheless, there was one more epoch-making aircraft to come. From as early as 1954 the Royal Aircraft Establishment in Farnborough had been looking into the feasibility of a supersonic transport aircraft (SST). Since the Second World War, British aircraft manufacturers, including de Havilland, had been experimenting with swept wing, high subsonic speed aircraft using data gained from the Germans in 1945. The DH108 had broken the sound barrier in 1949, although the second prototype had crashed tragically killing its test pilot, Sir Geoffrey de Havilland's son.

By 1956 the British government was taking the idea of SSTs seriously, and in 1959 contracts for preliminary designs were awarded to Bristol Aircraft and Hawker Siddeley. The French government

*Despite the sticker, Concorde was never seen in BOAC livery.*

*Concorde entered commercial service with British Airways in 1975, a year after BOAC had ceased to exist.*

had awarded similar contracts to Sud Aviation, Nord Aviation and Dassault. The British government felt that signing a deal with Sud Aviation would smooth Britain's entry into the Common Market (although General de Gaulle ultimately said 'non'). A draft treaty was agreed in 1962 which specified that the two countries would combine their resources to produce an SST which was to be called *Concorde*.

On 3 June 1963 BOAC, Air France and Pan American each signed an order for six Concordes with options for two more. More orders from other airlines followed but the original programme cost estimate of £70 million ran into huge overruns and delays, with the total cost eventually spiralling to £1.3 billion. By 1975 all of the sixteen airlines who had placed orders had cancelled, with the sole exceptions of BOAC and Air France. In truth, they would probably have cancelled as well, if their respective governments had allowed them to.

Concorde became a political football with an endless media and political debate raging over its cost, technical feasibility, and effect on the environment. BOAC, as a wholly government-owned corporation, had no option but to continue with the programme. Although BOAC pilots took part in route-proving flights and test flew the aircraft at supersonic speeds, Concorde came into service too late to ever appear in BOAC's livery. The first aircraft, registered G-BOAC, entered commercial service with British Airways in 1975, a year after BOAC had ceased to exist.

# CHAPTER THIRTEEN

## The end of The Golden Age

Politics would also play a role in the end of the Golden Age. On 6 September 1970, members of the Popular Front for the Liberation of Palestine (PFLP) hijacked several airliners. Two Boeing 707s, belonging to TWA and Swissair, were forced to land at Dawson's Field, a remote desert airstrip near Zarqa, Jordan (formerly Royal Air Force Station Zerqa), which then became PFLP's 'Revolutionary Airport'. On the same day, the hijacking of El Al Flight 219 from Amsterdam (another 707) was foiled: hijacker Patrick Argüello was shot and killed, and his partner Leila Khaled was overpowered and turned over to British police in London. On 9 September BOAC Flight 775, a VC10 operating from Bahrain, was hijacked by a PFLP sympathizer and brought to Dawson's Field in order to pressure the British government to free Khaled.

For several days British television viewers watched the hostages including children, who were held in conditions of insufferable heat and without any proper sanitation, being used as negotiating pawns by the terrorists. Finally, Leila Khaled was freed, and the passengers and crews were released unharmed. But the Dawson's Field hijackings had changed perceptions of flying – the public had long accepted the remote risk of an accident, but the idea of being held hostage, or killed by the deliberate act of terrorists, was an altogether different order of danger. Commercial airliners were now moving targets in a new kind of warfare which has continued up until the present day. Not only would this deglamourise flying, it would lead, in time, to the ultra-strict levels of security that we all now endure at airports.

# CHAPTER FOURTEEN

## The Merger

*'If this business were split up, I would give you the land and bricks and mortar, and I would take the brands and trademarks, and I would fare better than you.'*

—John Stuart, *Chairman of the Quaker Oats Company*

The decision by Atlee's government, in 1946, to create two new state-owned airlines, BSAA and BEA, while allowing BOAC to concentrate on its traditional Empire and transatlantic routes, now seems difficult to defend. It came as part of a huge post-war nationalisation programme in which civil aviation was one of many industries to be taken into public ownership. Today, virtually all of them, including civil aviation, have been re-privatised, but nationalisation remains a hot topic of debate. In the 2017 general election the Labour Party campaigned to bring rail companies, energy supply networks, water systems and mail delivery back into public ownership.

There had been talk of merging BOAC and BEA as early as 1953, following squabbles between the two over traffic rights. BOAC held rights to stop in several European cities, notably Rome, Frankfurt and Zürich, in order to pick up passengers for their Empire routes, but the fact that BOAC could also carry passengers into these cities from London had long riled BEA. At the same time, BEA had been using an agreement with Cyprus Airways to carry passengers beyond Europe into the Persian Gulf, which BOAC saw as exclusively their territory. Each accused the other of poaching and the row escalated to ministerial level. The chairman of BOAC, Miles Thomas, proposed a merger and got backing from the Chancellor of the Exchequer, Rab Butler, but Treasury mandarins poured cold water on the idea.

*A British Airways Boeing 747, beautifully repainted in BOAC's most famous livery (a golden Speedbird on a dark blue tail), is thrilling crowds at air shows.*

In 1969 the government commissioned the Edwards Report which recommended creating a new British Airways Board to manage both BOAC and BEA, as well as the two regional British airlines: Cambrian Airways (based at Cardiff), and Northeast Airlines (based at Newcastle upon Tyne). The report clearly recommended that: 'BOAC and BEA should retain their individual identities. There should be safeguards to avoid over-centralisation in the Holdings Board, a majority of whose members would also be on the boards of one of the Corporations or BAS.'

The new board took over in April 1972, and initially followed the Edwards Report's stipulation that each airline should retain its own individual branding, but within two years they had decided to fully merge the four carriers into a single airline to be known as British

Airways. The new airline commenced operations on 31 March 1974. BOAC, arguably the most recognised airline brand in the world, had disappeared, and it would take more than a decade for the new British Airways to establish itself as a clearly defined brand in the minds of the travelling public. By 1991, Pan American, the world's other great airline brand, had collapsed into insolvency - the Golden Age of Flying was over.

Today, the value of brands and corporate identities is better understood than it was in 1974. Analysts have shown that a popular brand can constitute 50 per cent or more of the value of a business. In 1989, the London Stock Exchange endorsed the concept of brand valuation by allowing the inclusion of intangible assets when seeking shareholder approval in acquisitions. Financial managers and planners are increasingly using brand equity tracking models for their analyses of businesses and calculations of their worth.

The airline business too, has lately come to recognise the value of well-established brands. When Air France and KLM were merged in 2004, each retained its individual identity. When British Airways' holding company, the International Airlines Group, took control of other flag-carrying airlines, including Iberia and Aer Lingus, they allowed them to keep their unique identities. Sadly, these lessons came too late for BOAC, BEA, Cambrian and Northeast, who were consigned to the archives of history.

Nevertheless, the brand proved too strong to vanish entirely without trace. As I write these words a British Airways Boeing 747-400, beautifully repainted in BOAC's most famous livery (a golden Speedbird on a dark blue tail), is thrilling crowds at air shows and reminding us that air travel once had a Golden Age that we shall never see repeated.

*BOAC's livery on a contemporary rally car (Photo Philippa Windsor).*

# *Index*